Electrical Power Generation Methods and Plants

T0271339

Electrical Power Generation Methods and Plants

Dhruba J. Syam

CRC Press
Taylor & Francis Group
Boca Raton London New York

CRC Press is an imprint of the
Taylor & Francis Group, an **informa** business

First published 2023
by CRC Press
4 Park Square, Milton Park, Abingdon, Oxon, OX14 4RN

and by CRC Press
6000 Broken Sound Parkway NW, Suite 300, Boca Raton, FL 33487-2742

CRC Press is an imprint of Informa UK Limited

British Library Cataloguing-in-Publication Data
A catalogue record for this book is available from the British Library

Library of Congress Cataloging-in-Publication Data
A catalog record has been requested

ISBN: 9781032516141 (hbk)
ISBN: 9781032516165 (pbk)
ISBN: 9781003403128 (ebk)

DOI: 10.4324/9781003403128

Typeset in Times New Roman
by Manakin Press, Delhi

Manakin
PRESS

भरतीय प्रौद्योगिकी संस्थान रुड़की

रुड़की 247667, उत्तराखण्ड, भारत

Indian Institute of Technology Roorkee

Roorkee-247 667, Uttarakhand, India

Tel.   : +91-1332-285597 (O)
         +91-1332-285038 (R)
Mob. : +91-9837074284
e-mail : spgfee@gmail.com
         spgfee@iitr.ac.in

FOREWORD

It was a great pleasure reading through the manuscript of "Electrical Power Generation - Methods and Plants" authored by eminent Shri Dhruba J Syam. Written in a masterly manner, it is a story of the current scenario of electrical power generation. Though continuous R & D efforts are being made, it still remains a challenge to generate eco-friendly power to meet ever increasing demand. For India, it is said, the future lies in developing technology to harness solar energy and to develop Thorium based nuclear power and also raise the per capita power consumption from the present 1000 Kwhr to 5000 Kwhr to be at par with the developed nations. Shri Syam has touched upon all issues related to power generation which makes his book an invaluable text for graduating engineers particularly in electrical and mechanical engineering disciplines as well as all those who are concerned with generation of electrical power.

S. P. Gupta

Dr. S. P. Gupta
Emeritus Fellow (Professor)
Department of Electrical Engineering
(Formerly Deputy Director)
Indian Institute of Technology, Roorkee
India

Acknowledgement

The Author is thankful to Sri. Vijesh Jain, Addi. General Manager (Electrical Machines Engineering) and Sri. D.K. Ray, Dy General Manager (Steam Turbine Engineering),- both of BHEL-Hardwar, for providing guidance in the preparation of the chapter relative to Thermal Power Plants. The Author is also thankful to Sri. R.P. Goyal, Retired Addl. Manager of BHEL-Hardwar, for his valuable suggestions on the chapter on Hydro-Power. The Author expresses his gratitude to Dr. S.P. Gupta, Professor-Emeritus in The Department of Electrical Engineering of the Indian Institute of Technology, Roorkee, for going through the manuscript and giving a Recommendatory FOREWORD for the book.

The Author is also thankful to M/s. Manakin Press Pvt. Ltd., New Delhi, for accepting and undertaking the printing and publication of the book.

—**Author**

Preface

With the increasing levels of Energy requirements and consumption arising out of population growths and aspirations for better quality of life, the demand for Electrical Power as the most convenient source of energy supply, will go on increasing. The Electrical Power Generation Methods and Plants and their service delivery capability and quality assume critical dimensions in the pursuit of overall developmental efforts of the Nations.

In the modern era, per capita consumption of electricity in a country is considered as one of the important indices of it's developmental status: both economic and technological.

In view of the above, logically, a country has to give due priority and make efforts towards development and availability of adequate Electrical Power Generation capacity on a sustainable regimen.

Engineering Students (electrical and mechanical in particular) as well as the professional beginners, studying and working in the field of Electrical Power Generation and Power Plant Administration, should get a reasonable level of familiarization with the concepts of various technological methods and plants in order to acquire necessary knowledge and competency for a worthwhile professional career in the subject field.

This book attempts to provide certain relevant knowledge inputs to supplement the Academic inputs, by way of providing conceptual clarity on various aspects of the subject. The Author hopes that the readers will find the contents educative and professionally useful.

The readers may refer to the list of books at ANNEXURE-VI for supplementary reading.

— **Author**

Brief Contents

Detailed Contents

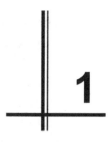

Electrical Power: A Convenient Source of Energy Supply

> *"Oh Varuna! Bring forth showers and vajragni, the sources of shakti that my creations need for sustenance, but avoid destruction while doing so"*
> ...... *Brahmadev to Varuna, the God of the Climate*

1.1 THE GENESIS

(A) As per the scientific theories so far advanced, the whole of the Universe (the galaxies, the planets, stars, other celestial bodies, the black holes, even the unfathomed voids) got created as a result of an explosive incident (the 'Big Bang' theory), and subsequent tumultuous celestial happenings, eventually leading to the liberation of Energy in immeasurably high quantum. This liberated energy assuming different forms, pervaded all over the universal domains including the Milky way, the Galaxy where our dear earth got created as part of the solar system and became resident therein.

(B) The share of the liberated energy that eventually got associated, directly and indirectly, with the Earth through the solar system, enabled it to have its own atmospheric sphere. This in turn, led to the creation of various features, both the biologically living and the non¬living , on the earth surfaces and sub-surfaces.

(C) As part of the process of evolution, human race also arrived on the scene along with or subsequent to the appearance of plants and animals as a result of the nature's proliferation processes.

Human beings because of their higher levels of intelligence, started recognizing the necessity of the energy, in various forms, to sustain their livelihood and their struggle for survival. With the passage of time, over the millennium, human race grew in numbers, needing increasing quantum of energy inputs from foods and other natural resources, not forgetting the **Lord of the Ring, the Sun.**

(D) With industrial revolution started by later part of eighteenth century/ early nineteenth century, need for energy started growing manifolds, and this trend continues in the present time also because of:

    (*i*)   Continuously growing world population aspiring for better standards of living, leading to ever increasing rate of consumptions for various purposes; this also includes energy requirements towards the live-stocks (domesticated animals adopted by the humans to support their living), which also grew in number and shared the available sources of energy supply thus boosting the level of energy consumption in various ways,

    (*ii*)   Ever increasing pace of industrialization and agricultural activities,

    (*iii*)   In spite of the best of international intentions and efforts to control and avoid, the increasing pace of militarization in different regions also contributes to the increasing levels of energy consumptions.

(E) Several Epoch making scientific and technological discoveries and Inventions during the last three centuries, prompted the human societies also to discover/ Invent various sources of energy to meet growing needs. One of the major sets of inventions towards this end, was the development of processes of generating and using Electrical Energy, now being put to use for almost all conceivable human endeavours, both industrial and non-industrial.

(F) **Power-Energy Relationship:**

    (*i*)   Using some notable scientific theories/concepts and inventions, developed and propagated during the last two centuries as listed below, it became possible to generate, transmit, distribute Electrical Power in large quantum to various load/consumption centres, to be eventually used as

Electrical Energy (E_e); the relationship between Electrical Power (P_e) and Electrical Energy (E_e) in a Direct Current (DC) system can be represented as under:

$E_e = P_e \times t$, where 't' is the time period of consumption of P_e for any application ;

Universally, E_e is expressed in kwhr (kilowatt-hour), the unit of electrical energy.

(ii) Since in the present era (starting approximately from the middle of the 20th Century), Electrical Power Supply System universally adopted the Alternating Current (AC) mode, for certain distinct tecno-economic advantages, specially at the users' end, besides those associated with the evacuation of the generated power through transmission/distribution systems. Power-Energy Relationship in the AC System can be represented as under:

$E_{eac} = P_{ac} \times t$ = Energy consumed in AC System over a period of time 't',

where, $P_{ac} = VI \cos \phi$ (= power in AC System),

where, V = system voltage. I = current flow in the system;

$\cos \phi$ = system power factor (generally recommended value is around 0.85 lagging minimum).#

\# **NOTE:** This will depend on the magnitude of Inductive and Capacitance current loads In the system,-both due to the internal Impedance of the system, as well as due to the nature of the externally connected electrical loads.

(iii) **Some of the major scientific theories/concepts and inventions, developed and brought into practice with specific reference to Electrical Power generation, can be listed as under:**

- Theories of Electro-Magnetism;
- Coulomb's Law;
- Ohm's Law
- Lenz's Law;
- Faraday's Law of Electro-Magnetic Induction,
- Electro-Chemical (Electrolysis) Process for Electricity Generation (Voltaic Cells).

- Theories of Thermo-Dynamics leading to development of Heat Engines such as: Steam Engines, Diesel Engines, Steam and Gas Turbines; to be used as Prime-Movers for Electrical Generators.

- Harnessing the Kinetic Energy of the falling waters (*), to be converted to Electrical Power/Energy through running of Hydraulic Turbines driving matching generators.

- The Potential Energy of the water-mass held in the Reservoir behind the Dam/Barrage, gets converted into Kinetic Energy during the flow down (through the penstock) into the hydroturbine blades, and in turn getting converted into rotational mechanical energy in the form of Torque delivery to the coupled generator rotor shaft.

- Nuclear Fission based Reactors to generate Heat to be used for generating Steam (through HRSG) to drive Steam turbine Generator set to deliver electrical power.

- Technologies for harnessing Solar Energy for Electricity generation.

- Technologies for harnessing Wind Energy through development of wind turbine driven generator.

- Technologies for Fuel Cells (hydrogen-oxygen gas interactive/reverse electrolytic reaction process based) **(See Chapter 7 for More on This Topic)**.

- Technologies for Geo-Thermal Energy: Hot water/steam/ hot fluids emanating from Hot-Springs/Geyser spouting from deep under ground, which can be used through HRSGs to run smaller Steam Turbine-Generator sets for vicinity area load-centres. Bore Holes drilled deep into the earth surface (into, even below the tectonic plates, at certain locations can bring out hot water, steam, hot gases/fluids to the earth surface to utilize the heat energy to generate electrical power **(See Refer to Chapter 7 for More on This Topic)**.

- Technology for harnessing the energy of Tidal Waves of the seas to run capsule/sub-merged Hydro-generating sets.

1.2 STATUS OF ELECTRIC POWER GENERATION CAPACITY BUILD UP IN INDIA AND OTHER COUNTRIES

In the present day world, one of the important indices of National Economic Development and standard of living in a country, is the per capita electrical energy consumption; at **ANNEXURE-1**, we are presenting a Table showing data on population, installed capacity of electrical power generation and the per capita electrical energy consumption of various countries (there are around 196 countries: big and small). From this table, we can see that in spite of substantial industrial and economic developments during the last six decades or so, per capita electrical energy consumption in India is only around 1000 kwhrs annually which is rather low when compared to those in some of the neighbouring Asian countries.

The lower rate of consumption in India can be attributed to the following main reasons:

(*i*) Wide gap in growth of installed capacity vis-a-vis growth of population: from around 1300 MWs for a population 430 Millions in 1950 to around 333550 MWs for a population of around 1270 Millions in 2017 (Ref: Annual Report-2017-18 and the website of the Ministry of Power, Govt, of India).

(*ii*) A vast area of the country not yet covered by Electricity

Network: the present National Target is to achieve this by 2022:

- Increasing the installed capacity to over 3,50,000 MWs;
- Near 100% coverage of all regions.

(*iii*) A substantial quantum of electrical power is lost due to inadequacy in transmission and distribution system networks (commonly known as 'line loss'); the national Government is taking steps to improve the situation by strengthening transmission and distribution net-work systems; one of the option being actively considered and adopted is to install an elaborate Net-work of HVDC Transmission Lines, specially for long distance transmission at 500 KV DC/750 KV DC or even 1100 KV DC to :

- Reduce the size (diameter) of conductors since at that thigh voltage, the magnitude of current flow is much lower; thus reducing cost on conductors also (which is a major component of over all cost of transmission);

- Avoid line loss due to **corona effect** around the Transmission line conductors at EHV AC transmission.

  [HVDC = High Voltage Direct Current;

  EHV = Extra High Voltage];

(*iv*) Low utilization and purchasing capacity for electrical power/ energy by over 30% of the population being in the BPL category; this situation is likely to continue for quite a number of years (may be over another decade, with respect to 2016); the situation is expected to improve gradually matching with the higher standards of living to be achieved for the BPL families (a long drawn-out affair in spite enhanced National priorities and efforts in this sector.).

REMARK:

HVDC Transmission System is a costly option: both at the sending end as well as at the receiving end. An elaborate Cost-Benefit Analysis has to be done to determine the extent of Trade-Offs between EHV AC system and HVDC System that can be achieved considering:

- The quantum of power to be evacuated/transmitted;
- The length of the transmission line to be erected; and the system voltage to be adopted;
- The maintenance and servicing costs involved;
- Quantum of line loss in EHVAC system due to line impedance and corona effects, vis-a-vis the impedance loss in HVDC system.

REMARK: Besides the cost factors, HVDC Transmission System is technologically more stable and reliable.

1.3 CONCLUDING REMARKS

(*i*) Considering the presently available technology, electrical power, for that matter, electrical energy, is the most convenient and preferred form of energy, both for industrial and non-industrial purposes.

(*ii*) By and large, Electrical Energy is eco-friendly at the users' end;

(*iii*) Electrical energy can be made available in large quantum from a single location source;

(*iv*) Electrical energy is a cleaner option at the users' end;

(*v*) Electricity is available in different modes, but we are limiting our discussion in this book (in the Chapters that follow) to the topics on Commercial Electrical Power Generation Methods primarily for the benefit of the engineering students (BE/B.Tech/AMIE) particularly those in the Electrical and Mechanical Engineering Streams. **This book seeks to provide certain conceptual clarity on the subject to supplement the academic knowledge inputs.**

(*vi*) Please refer to the following for additional but relevant information:

- **Annexure II** for a brief on Energy Resources of India at a Glance.
- **Annexure III** for a list of Energy Terms and Definitions.
- **Annexure IV** for a brief on Quantities and Units often used in engineering and scientific applications.

2

Electrical Power Generation Methods and Associated Aspects

> *"Professional knowledge needs regular up-gradation and application to avoid being stale. Outdated knowledge is like carrying a bag of trash."*
>
> — *From a lecture delivered by an Academician.*

REMARK : *Electrical power can be generated in a number ways depending on the input sources and the technology adopted. In the paragraphs and Chapters that follow, we are dealing with the methods for commercial generation of electrical power. This book is not supposed to serve as Text Book, but as a help book for enabling understanding of the basic principles of various methods of electrical power generation. Elaborate engineering and design details are not being discussed here. Diagrams/Figures presented are supposed to represent schematic concepts. Engineering students, particularly those in Electrical and Mechanical Engineering Streams, are advised to read this book as a supplementary to the Academic knowledge inputs, thus enhance their conceptual understanding of the subject.*

2.1 Based on the current technological knowledge available and in practice, in the following Chapters, we are attempting to discuss various methods of Electrical Power Generation, as listed under;

2.1.1 Thermal Power Generation Plants

These generally refer to :

(a) Steam Turbine Driven Turbo-generation (STG);

9

(b) Gas Turbine Driven Turbo-generation (GTG);

(c) Diesel Engine Driven Generation (DG);

(d) Combined Cycle Power Plants (CCPP and IGCCPP);

  * **NOTE:** Steam Engine driven generation system is now an obsolete concept and not in use anymore for commercial power generation.

2.1.2 Hydro Power Generation Plants

(a) Kaplan Turbine Driven;

(b) Francis Turbine Driven;

(c) Pelton Wheel Driven;

(d) Reversible Turbine Driven;

(e) Bulb Type Hydro Sets;

2.1.3 Nuclear Power Generation Plants (Commonly Adopted for Commercial Power Generation)

(a) Heavy Water Reactor Based;

(b) Light Water/Boiling Water Reactor Based;

(c) Gas Cooled Reactor Based;

(d) Fast-Breeder Reactor Based;

2.1.4 Solar Energy Based Power Generation.

2.1.5 Wind Turbine Based Power Generation.

2.1.6 Co-Generation Power Plants;

2.1.7 Bio-Mass Based Power Plants;

2.1.8 Geo-Thermal Power Plants;

2.1.9 Tidal Energy Based Power Plants;

2.1.10 Fuel Cells Based Power Generation System.

2.2 TECHNO-ECONOMIC ISSUES RELATED TO ELECTRICAL POWER GENERATION METHODS

REMARK:

Before we go into the Technological details of various Methods (paragraphs 2.1.1 to 2.1.10 above), in the following paragraphs, we are discussing the various Techno-Economic issues that get *associated with the listed power generation methods to enable the readers to appreciate the over all scenarios with respect to commercial generation of electrical power. We believe concentrating only on technical aspects leaves the discussion half-way. As future Business Administrators, young engineers should have some idea/ familiarization with the issues concerning the **power business management**.*

2.2.1   Referring to all the Generation Methods Listed above under paragraph 2.1, while the fundamental scientific concepts (Laws of Physics in particular) apply to each case, there are variations in configuration, construction and operational technologies; therefore, it may be useful to deal with the techno-economic issues pertaining to each, as described under, before going into the technical aspects:

(A) In the Case of Thermal Plants (STG and GTG)

(A-l)   **Fuels:** STG and GTG plants are based on the use of Fossil Fuels (coal, lignite, and petroleum derivatives such as furnace oil, natural gas, naptha, diesel);

The selection of the fuel for STG units is dependent on:

(*i*)   Steam turbine design philosophy and technology proposed to be adopted; sizing and configuration of the turbine will be a critical issue, both from the point of view of desired optimal output capacity as well as operational efficiency with respect to fuel consumption and heat rate achievable under different loading conditions.

(*ii*)   Related techno-economic issues: comparative advantages for cost-effective operations;

(*iii*)   Location related constraints for fuel supply and availability on regular sustainable basis, as well as associated costs towards transportation/logistics, storage, handling.

(*iv*)   The capacity range of the units and the total capacity proposed for the station (number of units to be installed).

(*v*)   The quality of fuel (calorific value, heat rate, contamination level: coal and lignite can have ash and shale contents higher than permissible limits, thus requiring coal benefaction, washery process before being supplied to STG plants; this will push up cost of fuel).

(*vi*) Fuel linkage (with suppliers) for assured supply on a regular basis for sustainable operations.

The selection of fuels for GTG Units is Dependent on:

(*i*)   Type of gas turbine and the output rating;

(*ii*) Duty-cycle being planned for the unit;

(*iii*) Techno-economic issues involved in operating the unit;

(*iv*) Whether an open-cycle or a combined-cycle operating system;

(*v*)   Any other relevant aspect unique to the proposed location;

(A-2) Other Relevant Issues to be Considered

(*i*)   Estimated Load demand pattern and Plant Load Factors will be important considerations for selecting unit size and the number of units to be installed in the proposed STG and GTG power stations. This is decided by the experts while preparing the DPR (Detailed Project Report) for the proposed power station.

(*ii*) Selection of location and analysis of advantages, disadvantages.

(*iii*) Capital Investment required, capital structure, ownership arrangement and the funding method to be adopted; provision for contingency fund for accidents, fire incidence, earthquakes, wars etc., to be made;

(*iv*) Life cycle of the selected plant and the depreciation rate to be charged.

(*v*)   Estimated rate of inflation (and cost escalation): during construction, during regular operation.

(*vi*) Taxes and other levies imposed by the Central and State Governments, local bodies;

(*vii*) Interest rates applicable: for capital funding as well as for working capital funding.

(*viii*) Expected concessions and subsidies available from the Central and the State Governments.

(*ix*) Cost of land and quantum of compensation that may have to be paid to the land evacuees/land sufferers.

(*x*)   Estimated Logistics and Transportation Costs: during project Construction, during regular operation.

(*xi*) Cost of construction of the power plant; costs towards project site office, Site stores and storage yards.

(*xii*) cost towards security arrangements.

(*xiii*) Total operating costs of regular operations taking into consideration:

    (*a*) Plant Load Factors/demand pattern.

    (*b*) Fuel costs and other associated incidental costs.

    (*c*) Costs of salaries and wages, other fixed costs.

    (*d*) All other incidental expenses associated with operations.

    (*e*) Interest rates and depreciation charges to be considered.

(*xiv*) Any other related operating costs unique to the subject project for arriving at the cost of generation per unit (kwhr).

(*xv*) Tariff system to be adopted (2-tier; 3-tier, fixed part, variable part).

(*xvi*) Any other technical and economic issues not mentioned above but are unique to the project.

(*xvii*) Determining the Break-Even period: Estimated and Acceptable to the promoters and funding agencies.

(*xviii*) Rate of Return on Capital investment: estimated and acceptable to the investors, loan givers (Banks/FIs).

(B) In Respect of Hydro-Power Plants

(*i*) Application of technology and design/construction features will depend on water flow conditions (rate of discharge, head available; perennial and/or seasonal).

(*ii*) 'Run of the River' project, or a dam and reservoir to be constructed/created.

(*iii*) Maximum head which can be created.

(*iv*) Up-stream and down-stream facilities required and to be created.

(*v*) Capital investments required, funding sources, capital structure to be adopted, cost of capital; ownership pattern to be adopted.

(*vi*) Operating costs, cost of generation per unit, taking into consideration, Plant Load Factor, Plant Availability Factor, demand (loading) pattern expected.

(*vii*) Break-even point for the project.

(*viii*) Estimated plant life-cycle and rate of depreciation to be charged.

(*ix*) Quantum of compensation to be paid to land sufferers/ population to be evacuated.

(*x*) Availability of construction power and skilled Manpower at the project site.

(*xi*) Cost of site offices and site stores.

(*xii*) Taxes and levies to be paid to the Central government and State Agencies, local bodies.

(*xiii*) Security related costs.

(*xiv*) Quantum of contingency fund provisioning to be made for accidents/mishaps (fire hazards, flooding, earthquakes, war, prolonged strike by employees).

(*xv*) Tariff scheme to be adopted.

(*xvi*) Manpower required: during construction, during regular operation.

(*xvii*) Costs towards project site offices, stores.

(*xviii*) Insurance coverage costs.

(*xix*) Overall cost of operation vis-a-vis expected Plant Load Factors and demand /loading patterns (per unit generation).

(*xx*) Who will bear the cost of power evacuation network to be created for connection to the nearest grid system? what will be the cost sharing arrangement for the same.

(*xxi*) Any other issue not mentioned above but is unique to the project.

(C) In Case of Nuclear Power Generating Plants (NPG)
REMARKS:

(*i*) In the Indian context, so far (2018) the National Policy is to have all nuclear plants under NPC (Nuclear Power Corporation, a Central Govt. Undertaking); therefore, many of the economic issues mentioned above for thermal and hydro plants will not be applicable in case of Nuclear Plants; the entire capital investment is borne by the Central Government; the operating costs, whatever these are, being met by NPC it self with some budgetary support from the Central Government,

(*ii*) For nuclear power projects, the capital costs and the operating costs data are not readily available since NPC will like to keep such information restricted to certain selected authorized agencies (mostly government), may be for certain valid reasons. However, for such heavy capital investments (estimated to be ranging from ₹ 6.5 crores to 7.5 crores per MW capacity, depending on the size, technology, and the location selected), NPC must be doing the Cost-Benefit Analysis while preparing the DPR (Detailed Project Report) to get government approval, before getting Budgetary Support; part of the funding is obtained from Banks/FIs as required.

(C-1) Techno-Economic Issues for Nuclear Plants

Technology for such nuclear plants are not readily available because of Dual-use restrictions imposed by technology givers: only a few industrially and technologically advanced countries can offer such technology and equipment (specially Reactors and certain associated auxiliaries). Over the last three decades, Indian organizations such as NPC, BHEL, L&T have mastered the technology of fabrication of some Reactor parts/components, certain auxiliary equipment, and almost 100% capability for manufacturing of associated HRSG, turbine-generator sets (up to around 700 MW unit capacity, mainly by BHEL). However, certain high technology restricted items (coolant pumps, fuel loading and un-loading systems, reactor control systems) even now have to be obtained from foreign sources (so far from Russia only). Recently, two Russian supplied 1000 MW units have been commissioned; for four more such 1000 MW sets, agreement has been signed with Russia in October 2016; these sets may become operational in the course next 3-4 years.

REMARK: Proposals are under active consideration to Procure and install 4-6 sets of 1000 MW capacity from Westinghouse, USA, in the course of next 5-6 years.

(C-2) Constraints in Nuclear Fuel Supply

India has very limited deposits of a radio-active fissile materials (Uranium) and is heavily dependent on supplies from some friendly countries such as Canada, Russia, France, Australia, Japan, to mention a few prominent ones. With India's entry into NSG as an associate member in 2008 (Full membership not yet conferred mainly because of Chinese non-cooperation), opportunities have now opened up for obtaining nuclear fissile materials (U238, U235) from some friendly countries. Japan has recently agreed to help in this regards. Because India has not yet signed the NPT (Nuclear Non-proliferation Treaty), some other NSG countries are not yet inclined to help.

As against the present (Dec. 2018) Nuclear Power Generation (NPG) capacity of around 6780 MWs, national objective is to achieve an installed NPG capacity of around 40000 MWs by next 10 years. With the shortage of nuclear fissile materials in the country, international restrictions on export of nuclear technology and materials, achieving the above mentioned target will be a challenging task; efforts are on for sorting out these bottlenecks with cooperation from Russia, USA, Japan, Australia, South Africa.

Please refer to ANNEXURE-5 for list of NSG countries and ANNEXURE-5(A) for Nuclear Power installed capacity in different countries (2016).

(D) In case of Diesel Generating (DG) Sets

(D-I) Techno-Economic Issues

(*i*) There are technological constraints in designing, operating and maintaining large DG sets of unit capacity above 6000 KWs; not normally recommended for regular operation (in continuous duty mode).

(*ii*) Fuel is costly and consumption rate for higher power output is rather high; so is for the lubricating oils to be used.

(*iii*) Not normally expected to be run continuously for long hours; mostly used for Emergency or Peak-Load back-up power supply.

(*iv*) Break-down rate is high in case the set is run continuously for long hours; rate of wear and tear on moving/rotating parts is high needing frequent replacements, a costly affair.

(*v*) DG sets below 15 KWs capacity is grossly un-economical in case run for long hours; recommended to be used only as Emergency power supply option.

(*vi*) Loading pattern is also a critical factor in regular operation.

(*vii*) Capital investment is much lower than those for STG and GTG plants of comparable size of out put; require much smaller premises and much less number of auxiliaries and needs only a few operators for each set/station.

>> **Because of mainly economical disadvantages, Installation DG sets are at a lower preference; total installed capacity in India (2018) is only 838 MWs. Saudi Arabia and Kuwait have DG Stations of large capacity because of lower fuel costs prevailing there.**

(E) In the Case of Solar Power System

There are two kinds of arrangements to capture heat and light from sun rays, for generating power:

(E-1) Parabolic Reflectors capturing heat from sun-rays and passing the same to water flowing inside pipe lines placed at the focal points in an array of such parabolic reflectors; heated water is then passed through insulated pipelines to an insulated reservoir/tank from where hot water is supplied to the consuming points through insulated pipelines. This method however, is not techno-economically suitable for commercial generation of electrical power in the present day context and therefore, not adopted.

(E-2) Solar Photo-Voltaic Panels

(This is the most commonly adopted technology now a days).

Converting captured sun-rays directly into Electricity.

The techno-Economic Issues that are worth mentioning are presented below:

(*i*)   For an appreciable quantum of electrical power generation, a large number of photo voltaic panels laid over a vast land area is required; this is a constraint for many load-centres; besides, capital cost can be substantial for a larger set up.

(*ii*) The technology of fabrication of photo voltaic cells (made of ultra-pure silicon grains/pastes) is quite intricate and is still evolving through R&D by many industrial enterprises; cost of development is also quite high considering the fact that the whole fabrication/manufacturing process has to take place under very high degree of controlled atmosphere (dust-free with lowest possible humidity and proper temperature control). Therefore, cost will be highly dependent on the quantum of finished products (cells) to come out during a cost-effective time period, and the rate of off-takes.

(*iii*)   Since the panels are mounted outdoor, cleaning and keeping them free from dust (and ice) deposits is a challenging and costly task.

(*iv*)   Availability of enough sun-rays during the day is also a critical factor; fortunately vast areas of India gets appreciable quantum of sun-lights through out the year.

(*v*)   Being a part of the National Objectives for power capacity development in the Renewable Sector, there is possibility of generating substantial quantum of electrical power through this system. According to the estimates made by the Ministry of New and Renewable Energy, Government of India (2015-16), there is potential in India for generation capacity of 7,49,000 MW but with present techno-economic limitations, it will require decades to harness a substantial part of this potential. The present (2018) installed capacity of solar power in India is around 17,000 MWs.

Private Sector participation in a big way can help achieving the target to a large extent; many parties are already in the field.

(F) In the Case of Wind Turbine Driven System:

(*i*)   In the last three decades or so, this method of power generation has emerged as a good option in the area of Renewable Energy Sources. Many countries including India have adopted/are going in for increasing capacity addition in this field.

(*ii*)   Technological developments undertaken during the last two decades or so, it has been possible to increase unit capacity from a few KWs to over 8000 KWs (in India, so far, maximum unit capacity installed is around 2000 KWs).

(*iii*) This is by and large, the cheapest option in the field of **Renewable Energy Sources**, enabling substantial quantum of power generation in one location provided:

• Enough wind flow rate is available over long hours;.

• Enough suitable land area is available for installation.*

*NOTE: Many countries in Europe have installed wind turbines in the coastal areas, even on the coastal continental shelves where good amount of wind is flowing most of the year.

(*iv*) The wind turbine set of UNIT capacity size of above 2000 KWs is NOT as simple as appears to be; the major technological problems to be tackled for:

- Balancing the equipment weighing more than a tonne, (in some cases, several tonnes) on a single slender tower (in some cases, over 150 metres high) under variable wind speeds and wind-force which may be quite high during storms/cyclones; besides incidence of occasional earthquakes also have to be taken into consideration.

- Speed control mechanism to achieve stable/uniform speed regimen under variable wind flow rates (speed is critical to achieving desired level of generator terminal voltage as well as the prescribed generation frequency).

- Carrying out periodical preventive maintenance and break-down maintenance are tricky jobs and require the services of highly skilled (and courageous) trouble shooters. As per estimates of the Ministry of New and Renewable Energy, Government of India, there is a potential in India for Wind Power Generation capacity of the order of 1,02,788 MWs; the present (2018) installed capacity is 32849 MWs. Plans are in hand to increase capacity substantially in the coming decades.

(G) In the Case of Fuel Cells

(*i*) This concept although more than a century old, is still evolving and not yet a technologically and economically viable option for large scale commercial power generation.

(*ii*) There are still technological limitations to be tackled for making the system suitable for larger scale generation and distribution to consuming centres. At present, trials are being conducted for vehicular use and for isolated smaller rural areas/sub-urban areas. Certain large ships are adopting this method of power supply (Please refer to Chapter 7 for more details on Fuel Cells).

2.3 ECOLOGICAL/ENVIRONMENTAL ISSUES

There are a number of issues that get associated with Power Plant operations. The relevant ones are listed out and discussed at the end of each Chapter on specific power generation plants detailed out in the Chapters that follow.

• Power Generation and the Environment ($)

The burning of fossil fuels e.g. coal, oil or gas inevitably causes local air pollution such as sulphur-dioxide (Sox), nitrous oxide (Nox) and dust particles. Modem power plants with **Electrostatic filters, (precipitators),** flue gas desulfurization plant and catalytic converters can considerably reduce such pollutants.

In contrast to these air pollutants, greenhouse gases such as carbon dioxide (CO_2), which are unavoidable in the process of burning fossil fuels, and are not possible to be retained/ neutralised by filters or catalytic converters, have no appreciable local effect but a global one. An efficient reduction of CO_2 therefore requires not only technical measures but also international agreements.

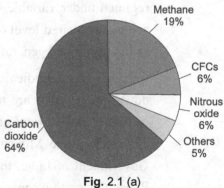

Fig. 2.1 (a)

Since the energy-sector (including traffic) produces approx. 50% of the greenhouse gases, measures to reduce CO_2-emmisions from power generation sources are becoming increasingly important.

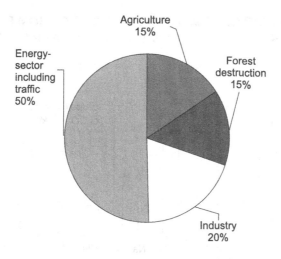

Fig. 2.1 (b)

If the installed worldwide coal-fired power plants were to be replaced by highly efficient modern plants, an annual reduction of approx. 1.6 billion tons of CO_2 emissions could be achieved. In the case of replacement by the combined-cycle plants, the annual CO_2 reduction would amount to around 4 billion tons.

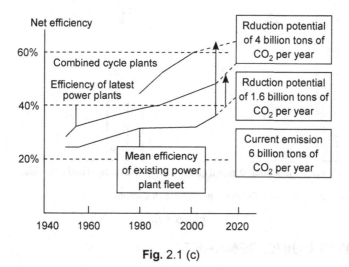

Fig. 2.1 (c)

CARBON-DIOXIDE EMISSIONS RESULTING FROM POWER GENERATION IN VARIOUS TYPES OF POWER PLANTS

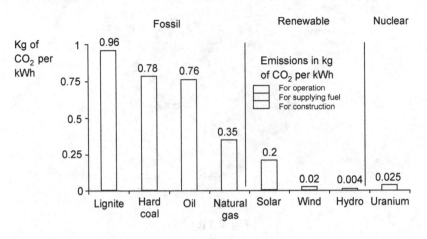

Fig. 2.1 (d)

POSSIBLE WAYS OF REDUCING CO_2 EMISSION (FOR OPERATION)

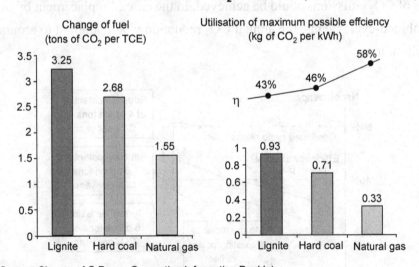

$ Source: Siemens AG Power Generation Information Booklet.

Fig. 2.1 (e)

2.4 CONCLUDING REMARKS

(*i*) The world (particularly the industrially advanced countries) is continuously on the look out for more dependable and economic options for large scale generation and availability of Energy Supply including Electrical Energy. With growing world population

(7.25 + Billions in 2017) and their aspirations for better quality of life, demand for energy is going to increase substantially in the coming decades.

(*ii*) International effort to develop a Controllable Fusion Reactor for achieving large scale Electrical power Generation, is a promising option but still several decades away to emerge as a practically feasible option. Once the technology is made available for practical applications, the energy availability problem is expected to be solved for ever since this Fusion Reactor will be using as fuel (fusion material) Deuterium (also known as Heavy Hydrogen,–an Isotope of Hydrogen) which remain dissolved in the sea water by natural process and is abundantly available. The planned Fusion Reactor, for higher energy yield, is proposed to use a mix of Deuterium and Tritium, another (Radio-active) Isotope of Hydrogen which is also found dissolved in sea water by natural process.

* (*iii*) Total Installed Electrical Power capacity in India (as on 31.12.2017) (Sector-wise) (MWs):

a. Thermal		b. Hydro = 44963	c. Nuclear = 6780
* Coal	= 1,92,972;	d .Renewables:	
* Gas	= 25,150;	* Wind	= 32,848;
* Diesel	838;	* Small hydro	= 4,418;
Total	= 2,18,960;	* Bio-mass/ Co-gen	= 8,419;
		* Solar	= 17,052;
		* Waste to Energy	= 114;
		Total	= 62,847

GRAND TOTAL OF ALL = 3,33,550 MWs.

*Source: Annual Report-and the Website of the Ministry of Power, Govt, of India.
Latest Figure can be Obtained from the website of CEA and Ministry of Power.

❑❑❑

3

Steam Turbine Driven Thermal Power Generation (STG) Plants

> *"A complex concept requires a complex approach in visualization, but the detailing should be done in simpler manner for enabling practical configuration and implementation"*
>
> — *Author*

3.1 INTRODUCTION

The STG Plants are designed to convert Heat Energy into Mechanical Rotating Energy, to eventually get further converted to Electrical Power which, at the user's end, provide the Electrical Energy for various usages.

3.2 A CONVENTIONAL STG PLANT

A conventional STG plant comprise of some major Equipment Systems and certain Auxiliary Equipment to make it work as a composite operating set up.

3.2.1 Major Equipment Comprise of

(A) **Boiler:** Generally water-tube type ones to generate super heated steam at the desired temperature, pressure and in required volumetric quantities (generally denoted in tonnes/hr).

(B) **Steam Turbine (ST):** To act as the Prime-Mover to provide the rotating mechanical force in the form of Torque to the coupled rotor of the Turbo-generator.

(C) **Turbo-Generator(TG):** To Generate Electrical Power when it's rotor acting as a rotating magnet (being excited with Direct Current in it's windings) creating a rotating magnetic field which induces Electro-Motive-Force (EMF) in the stator windings of the TG (by following the principle of Faraday's Law of Electro-Magnetic Induction); this process in turn creates a Potential Difference (also called Voltage) between different terminals of the Stator windings from where Power is drawn for the electrical load-networks. Besides the local distribution systems, the bulk of the generated power is evacuated through the Grid system (OH Transmission lines) to various load-centres located far and wide. All such distribution (HT and LT) and the transmission systems (HT/EHV AC, HVDC) work through the connected Transformers and matching Switch-gear/ circuit-breakers of appropriate capacity ratings with associated protection system (protection against over loading, fault conditions, loss of synchronism in certain circumstances).

(D) **Boiler House Complex:** Buildings, Structures, boiler combustion chambers with Fire-bricks lining, pipe-lines (A variety of them duly colour-coded; steam and hot water pipes duly insulated), fuel supply system, burners, other associated equipment and services.

(E) **Turbine Hall:** To house steam turbines, generators and their proximity auxiliaries, including local control panels, other local services.

3.2.2 Auxiliary Systems Comprising of:

(1) For the Boiler (Steam Generator):

(*i*) DM water supply system (DM = De-Mineralized) with dedicated Boiler Feed pumps, water treatment plant, associated pipelines, associated drives and controls.

(*ii*) De-Aerator system.

(*iii*) Water Reservoir/water source (Lake, River, Sea*).

(*iv*) Cooling water supply system with dedicated pumps, drives and controls.

(*v*) Steam flow pipe lines (heat insulated) with valves and control systems.

(*vi*) Steam blow-down system.

(*vii*) Boiler control panels.

(*viii*) ID (Induced Draught) and FD (Forced Draught) Fans and their drives and controls.

(*ix*) Assorted other small auxiliaries to support boiler operations.

* **NOTE:** In case sea water is used as primary source, then Desalinization plant will be required before the sea water can be used, specially for the DM water supply system.

(2) For the Steam Turbine:

(*i*) Steam inlet valves, Emergency Control/Stop valves, with dedicated drives and controls; steam pipelines (duly insulated) coming into the turbine cylinders.

(*ii*) HP and LP Heaters with associated pipelines and controls.

(*iii*) Turbine Speed Control Governing System.

(*iv*) Turbo-visory control systems and panels.

(*v*) Condenser with mounting pedestals, associated pipelines, valves and controls.

(*vi*) Cooling water supply system with associated pipelines, valves, controls.

(*vii*) Cooling Towers with Fans and drives, associated pipe lines, valves, pumps, controls.

(*viii*) Lubrication system for turbine rotor shaft bearings, with associated pumps, pipelines and controls.

(*ix*) Turbine Rotor Shaft sealing system (cylinder end-shield mounted).

(*x*) Turbine mounting bearing pedestals.

(*xi*) Turbine cylinder body insulation covering with Sheathings.

(*xii*) LP Bypass valves with associated pipelines and controls.

(*xiii*) HP Bypass valves with associated pipelines and controls.

(*xiv*) Boiler shut-off valves with associated pipelines and controls.

(*xv*) Barring Gear system (to be used at the time of initial starting of rotations of the rotors, as well as to be brought into operation when the ST is to be slowed down and stopped (in off-load condition).

(3) For Turbo-generator (TG):

$(*i*) Excitation system (for DC supply to the TG rotor windings to make it a rotating magnet).

$ **NOTE:** There are different methods for DC supply, but modern trend is to use high current rating Diodes fixed on the rotor for enabling Brush-Less injection of DC supply to the rotor windings.

(*ii*) Ventilation and cooling system (chilled air cooled, chilled hydrogen cooled, combination of chilled water-chilled hydrogen cooled), depending on the size/capacity ratings and the technology preferred;

(*iii*) Lubrication system for TG rotor shaft bearings:

(*iv*) Cooling (water or other fluids) system for the TG rotor bearing housings;

(*v*) TG rotor shaft sealing system embedded in the end shields.

(*vi*) TG control panels;

(*vii*) TG Transformer with associated Switchgears/circuit-breakers, bus-bars, bus-ducts, controls and protection system;

(*viii*) TG Earthing system (including the Earthing Transformer and associated controls);

(4) Station Auxiliaries (Assuming that the Power Plant is a Coal Fired One):

*(*i*) Coal yard, coal transportation, coal handling equipment, conveyor system, hoppers and chutes.

*(*ii*) Coal pulverization mills and coal dust spraying system;

*(*iii*) Ash and slag disposal system.

*(*iv*) Stack (Chimney) system.

*(*v*) Flue gas discharge system.

*(*vi*) Electro-Static Precipitators, gas scrubbers .

(*vii*) Station Switch Yard with structures, Transformers, Circuit-breakers, OH Bus Bar systems, CTs, PTs;

(*viii*) Station transformer for power supply to station auxiliaries with associated drives and controls; station lighting and yard lighting;

(*ix*) Station Central Control Room (normally adjacent to the Turbine Hall).

(*x*)   Station Earthing System and Lightning Arresters.

(*xi*)   Station Capacitor Banks (optional/as required).

(*xii*)   Station Communication Networks.

(*xiii*)   Station Road Systems, Rail-Yard, with loading and unloading bays and cranes (normally Gantry Type moving or rail).

* **NOTE:** Each STG unit will have its dedicated Boiler with Auxiliaries, ST and TG with their auxiliaries.

- **Figure No. 3.1** represents a Schematic Block Layout of a typical Coal Fired STG plant.

- **Figure No 3.2** represents the Schematic Arrangement of the Functional System of a Typical coal fired Steam Turbine-Generator set.

3.3 FUNCTIONAL FEATURES OF A WATER TUBE BOILER

3.3.1 Coal and Lignite fired boilers will have specially designed and configured fire-chambers for using pulverized coal/lignite dusts to be sprayed into the boiler fire chambers along with air (O_2). To fire up the boiler in the starting stage, flame throwers using a mixture of air and inflammable oil or combustible gas, ignited by electric sparks, initiates the flame in the fire chambers. Once the fire level in the fire chamber reaches a threshold level (it may take a few minutes for sustained flame up), starting sparking system is withdrawn into a recess away from the fire chamber, to be re-deployed at the time of re-start. The flame-throwers /burners are kept active as long as the fire-box is live with fuel burning process for continuous flame generation.

3.3.2 In case furnace oil or natural gas is used as the fuel, fuel injection along with air uses a different kind of system, with an appropriately designed ignition starting device to fire up the boiler. In this case, the systems of solid fuel handling, pulverization, mills, fuel conveyor belts, hoppers, chutes etc. are not required, but appropriate system for oil or gas supply into the boiler will have to be installed. In this case the coal yard is replaced by oil or gas (as the case may be) storage-tank yards with associated pipelines, pumps, with associated controls.

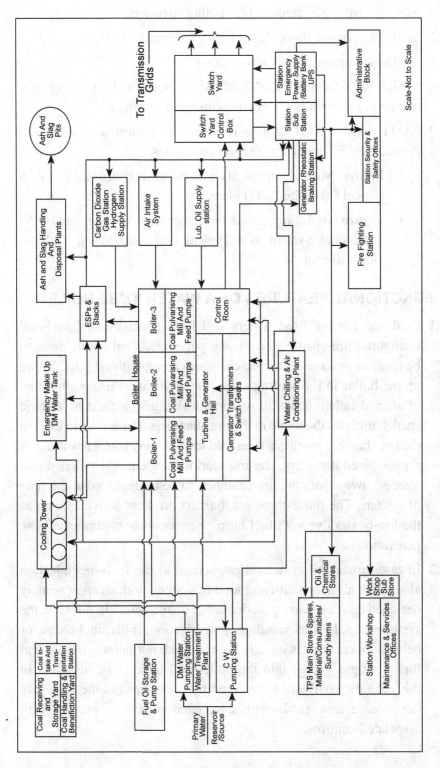

Fig. 3.1: Block-Diagram Layout Plan For A Typical Coal-Fired Thermal Power Station.

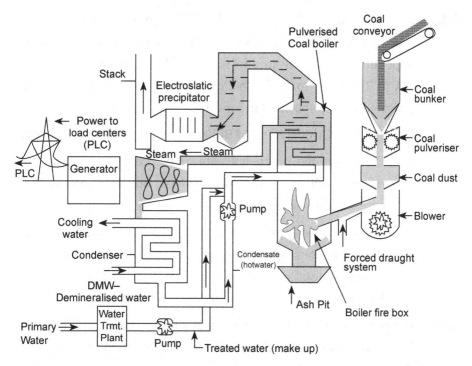

Fig. 3.2: Schematic Arrangement of a Conventional Thermal
Power Plant (Coal Fired)

NOTES:

(*i*) Duel Fuel System, for both solid or liquid/gaseous fuels, can be adopted if user so desire, but this has to be decided at the design stage of the boiler; this option will bring in added technological complexities and Cost of the boiler may go up substantially, besides involving higher maintenance costs.

(*ii*) Students of Power Plant systems should get familiarized with the significance and use of Steam Charts for design and fabrication of boilers/steam generators.

3.3.3 Modern Power Station boilers for Sub-Critical STG sets are designed to operate on continuous duty cycle to deliver super heated steam at 535°C to 545°C with pressure 150-170 ATA (ATA = Atmospheric Absolute).

3.3.4 With demand for Super-Critical ST design preferred for Unit capacity above 500/600 MWs, operating with superheated steam temperature ranging from 565°C to 630°C, at pressure from 246 ATA to 270 ATA; matching boilers also had to be developed and fabricated for the specified super-critical parameters.

NOTE: Super-Critical STs and Boilers are being preferred because:

(*i*) Better thermal efficiency with higher capacity units in the range of 600/660 MW, 700-750 MW, 800 MW, and in rare cases, 1000 MW as well

(*ii*) Larger capacity units are preferred for Super Thermal Power Stations so that numbers of units can be lesser for the aggregate total capacity planned for that power station; this has techno-economic advantages.

3.3.5 To operate with Super-Critical parameters, boilers will have to use special grade steel/alloy steel, titanium/tungsten alloy metals/ other super alloys that can safely withstand that level of thermo-mechanical stresses in the fire chambers structures as well as hot water as well as super-heated steam pipelines. Therefore, metallurgical aspects will come into play and the designers will have to get well conversant with such metallurgical aspects before selecting materials of construction.

3.3.6 'Once Through Boilers':

(A) For ensuing stream-lined operation with minimum of Line loss/ system loss with respect to heat and pressure, modern boilers adopt what is called 'Once Through' concept of steam generation and steam flow system, dispensing with Boiler Drums as the intermediate steam holding /accumulator vessels. Instead, the super heated steam is passed though appropriately placed 'Headers' which allows steam flow at the desired level to be directly fed (through the insulated steam pipe-lines) to the inlet valves of steam turbine. With such configurations, the overall sizes of the boiler house equipment and structures could be reduced appreciably (25-33%) and hence, substantially reduce the cost of the boiler house complex.

(B) Such boilers, for safe and satisfactory continuous operations, will require:

- Very close performance monitoring with digital indicators and instrumentation.

- Safety pre-cautions/back up systems at all stages of operation.

- Dependable materials of construction for critical components/ parts to safely withstand the thermal and mechanical stresses of that level.

- Strict adherence to prescribed periodical preventive checks and servicing/maintenance.

- Deployment of well trained operating and maintenance personnel.

- Adherence to the statutory provisions of the Boiler Act.

- Assessment of residual life for safe-operation after expiry of the recommended period of service years, and take measures for rehabilitation/ life-extension programmes through major overhauling.

(C) Boiler and the steam flow control systems will have to be Net-Worked and in tune with the control systems of the turbine and generator to take care of the variations in load demand at the generator terminals (through the generator transformer). This is particularly important when the generator trips i.e., Goes out of power delivery system under adverse operating conditions (system faults, system malfunctioning/generator-grid out of step/loss of synchronism, sudden load outage due to grid failure and the like).

3.3.7 While the fundamental working principles are similar, or same, Technological/Design variations can be there from Manufacturer to Manufacturer, depending on (for the same unit size of STG):

(*i*) Ambient conditions prevailing at the STG Plant location with respect to:

- Ambient temperature, atmospheric pressure, altitude of the location.

- Variations in ambient temperature from winter months to summer months.

- Cooling water ambient temperature.

- Ambient temperature of the DM water supply.

(*ii*)   Loading pattern expected throughout the life cycle of the STG unit.

(*iii*)   Grid connectivity and prevailing grid operating conditions.

(*iv*)   Type of Fuel to be used and fuel quality (calorific value per kg, contamination level expected in the fuel, particularly in coal/lignite which can have ash, shale, stone/silicon contents beyond permissible limits).

3.3.8 The entire ST set up, starting from the associated boiler and its auxiliaries, steam delivery pipelines, steam turbine and its auxiliaries, all associated Control Apparatuses and all other upstream and downstream facility including the coupled TG,

- All these are required to work in tandem to make it a composite working system for eventual generation of electrical power in a dependable/ assured operating regimen as recommended by the equipment manufacturers.

3.3.9 The Boiler, the ST and the TG, – all have to work in near perfect unison for a stabilized operation even under stressful situation of widely varying loading conditions.

3.4 FUNCTIONAL FEATURES OF A STEAM TURBINE

3.4.1 The steam turbine acts as a Prime-Mover and delivers rotating mechanical force (in the form of Torque) through its bladed rotors to the coupled rotor of the Turbo-generator (TG) (also refer para 3.2.2(3).

3.4.2 The super heated steam coming from the boiler (through insulated pipe lines) enters team turbine cylinders through the Steam Inlet Valves and impinge on the rows of fixed blades placed on the inner surface of the cylinders, as well as on the moving blades fixed on ST rotor shafts; the heat and pressure of the steam causes rotating mechanical motion (due to creation of reaction forces between the blades fixed on the inside of the cylinder and the rotor blades of the ST rotor), thus making the ST as a rotating machine.

3.4.3 For a conventional ST for sub-critical operations, the sizing and configuration will depend on:

(*i*) Unit size, such as : 100 MW, 200/210 MW, 250 MW, 270 MWs, 300 MW, 500 MW, 600/660 MWs etc.

(*ii*) Steam parameters such as: temperature, pressure, volume of flow, for different load conditions etc..

For Example: For a 500 MW unit (Siemens AG Technology based)

• Super heat temperature: 537°C at entry;

• Pressure (at entry) : 170 ATA.

3.4.4 Other Features:

(A) **Configurations:** numbers of cylinders and rotors adopted,

• For example: A 500 MW ST will have:

– 1 HP cylinder with bladed rotor: Single flow;

– 1 IP cylinder with bladed rotor: Double flow;

– 1 LP cylinder with bladed rotor: Double flow.

[HP = High Pressure]

[IP = Intermediate Pressure]

[LP = Low Pressure]

• **NOTES:**

(*i*) For sub-critical turbines, beyond 500 MW, LP cylinder is replicated twice or thrice depending on the unit capacity chosen;

(*ii*) For super critical STs, there can be single LP even up to 660 MWs unit capacity.

• **Figure No. 3.3** represents a set of schematic configurations of STs;

• **Figure No. 3.4** represents sectional views of STs of certain capacity ranges.

(B) **Steam Re-heat:** After working in HP cylinder, steam goes back to the boiler for a re-heat to the original super heat temperature; re-heated steam from the boiler comes back to the ST and enters the IP cylinder; such a system enhances the working capacity and the thermal efficiency of the turbine.

(C) **Steam Flow Paths:**

      (*i*)   Inside the turbine, in addition to the points in paragraph 3.4.2 above, the numbers of rows of blades and number of total blades inside the cylinders and on the rotors will very according to the unit capacity of the turbine and the configuration selected.

     (*ii*) One row of moving blades and one row of stationary blades together is called a stage,

    (*iii*) Number of stages in the cylinder will depend on how much pressure drop needs to be taking place in the cylinder.

            For Example: for a typical 500MW ST of sub-critical Siemens AG Design:

            – No. of stages in HPT : 17 stages;

            – No. of stages in IPT : 2×12 stages;

            – No. of stages in LPT : 2×6 stages;

     (*iv*) steam flow cross-over pipes from HPT to IPT and LPT can be as shown in the Figure No.3.4 for different ST capacity ranges.

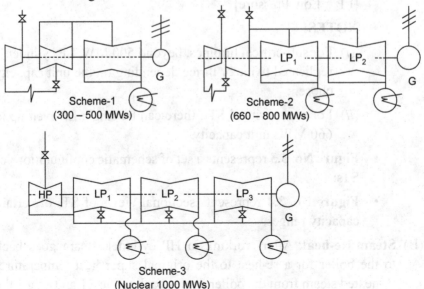

Fig. 3.3: Schematic diagrams of a few representative configurations of steam trubines for power plants *(Siemens AG Technology Based)*

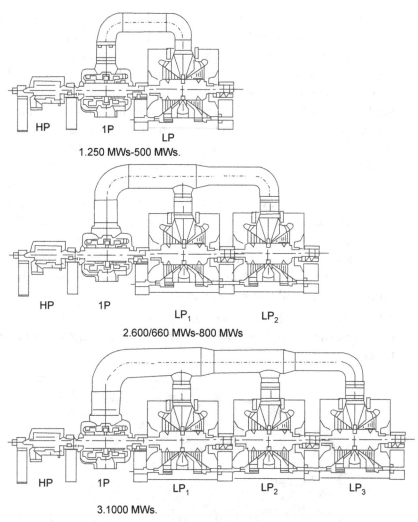

HP   1P
                        LP
1.250 MWs-500 MWs.

HP   1P
                    LP₁   LP₂
2.600/660 MWs-800 MWs

HP   1P   LP₁   LP₂   LP₃
3.1000 MWs.

Fig. 3.4: Schematic layout of steam turbines of various capacity ranges.
(Simens AG Techonolgy Based)

(D) **Governing system:** Rotating speed of the Turbine has to be controlled and governed according to the varying load conditions (on the coupled TG);

    Governing system continuously sense the torque-speed conditions of the turbine rotor and sends signals to the steam inlet valves (throttle governing principle) to allow ingress of more steam, OR, reduce flow of steam into the turbine, depending on the torque demand on it's rotor to match the loading trend on the coupled TG.

(E) **Condenser System:** After working on the blades of the last stages of LP cylinder, the steam becomes saturated with water droplets, temperature goes down appreciably, and the pressure goes down to near vacuum stage (0.3 Bar to 0.08 Bar depending on the ST design adopted); with this condition, the saturated steam passes over to the condenser placed below the LP turbine, where this steam gets condensed into hot water; this hot water is pumped into the hot well where from it is pumped into the boiler to again get converted into steam; this cycle goes on and on so long as the boiler is functional for delivery of super-heated steam to the turbine. In the hot water return circuit pipe lines, make up DM water is added, at an appropriate stage, into the water supply system. To enhance the thermal efficiency of the boiler, this water is passed through the HP Heater and LP Heater so as to pick up additional heat from there before the same is taken to the boiler water tubes. Both HP heater and LP heater, functioning as Heat-Exchangers, get bleed steam supply from the HP cylinder and LP cylinder respectively, by properly sealed and insulated pipeline lines.

The condensate hot water pipes also passed through the De-aerator to remove the air contents in hot water to avoid damage to the boiler tubes.

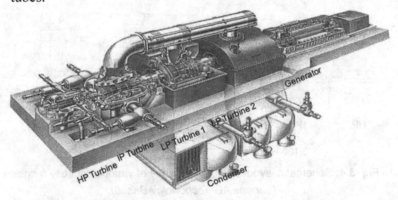

Fig. 3.5: Image of a large Steam Turbine driven Generator Set

NOTE: both HP and LP heaters have other functions also in the system by way of contributing to the thermo-dynamic regulatory balancing of steam flow inside the respective cylinders.

For larger STs having more than one LP cylinder, normally, each LP cylinder will have it's dedicated condenser, which gets connected to the Hot-well system of the dedicated boiler.

(F) **Metallurgical Characteristics:**

 • **For Cylinder Casings:** Creep resistant alloy steel explosion proof castings;

- **For Rotor Shafts:** Creep resistant alloy steel forgings;
- **For Blades:** Creep resistant alloy steel Precision forgings;
- **Steam Flow Control Valves:** Special grade steel castings (explosion proof), steel rolled stocks.
- **Appropriate heat Treatment for All Components, Parts, as Recommended by The Designers.**

(G) **Machining of Parts and Components:** As per the technological processes recommended by the manufacturing technologists.

(H) **Works Testing of ST:** While the bladed cylinders and rotors are assembled in the manufacturing plant to check the accuracy of assembly/any mismatch, the full turbine is steam tested only during the commissioning and eventual load/performance testing at the power station where the STG set is supposed to work.

- Prior to assembly, the bladed Rotors are subjected to Full Speed (10% – 25% over speed also) Dynamic Balancing in a Vacuum Tunnel equipped with drives, drive controls and vacuum pumps and associated controls.

Fig. 3.6: A Typical Thermal Power Station

$ (I) **Final Commissioning:** Done at the STG plant after erection and assembly works are satisfactorily completed.

$ **NOTE: With strict in-process Quality Control measures in place during manufacturing, there is rare chance of encountering any bottlenecks at the time of assembly and steam testing at the project site.**

REMARKS:

(*i*)   The user organization (of the STG set) may not like to go into the nitty-gritty of STG manufacturing processes and is most likely to concentrate on timely delivery, satisfactory erection and commissioning, performance testing and getting the Performance Guarantee Undertaking/Bond for specified period, from the Supplier/manufacturer,

(*ii*) However, some customer may depute their QC representatives at the manufacturer's works to monitor in-process quality control as per the prescribed Quality Plan as well as monitoring the manufacturing progress as per the agreed schedules..

3.4.5 Super-Critical Steam Turbines

(*i*)   As per modern trends, Power supply companies are preferring to install larger steam turbines with Super-Critical Parameters for STG sets in the unit capacity ranges above 500 MW, such as 600/660 MW, 700/750 MW, 800 MW and in rare cases, 1000 MW.

(*ii*) Super-Critical turbines are designed for operating with super heated steam at temperature from 565°C to 630°C at pressure ranging from 246 ATA to 270 ATA.

(*iii*) As already stated at paragraphs 3.3.4 and 3.3.5 above, in the case of super-critical turbines also, the same conditions will prevail.

REMARKS:

After the expiry of the recommended period of operating years, a full scale examination of the stressed parts have to be carried out to determine their residual life for safe functioning and essential rehabilitation/life extension programmes have to be taken up and executed in all due seriousness. Manufacturer is supposed to specify these aspects in the supply contract/Performance Guarantee Bond.

3.5 FUNCTIONAL FEATURES OF A TURBO-GENERATOR (TG)

(A) Asper the National Standards, the TG for the system is designed and configured as an Alternating Current (AC) 3-Phase 50 Hz, 3000 rpm machine (as in India) (or 60Hz/3600 rpm, depending on the National Grid System Preference).

Fig. 3.7: Photograph of a large turbo-generator

(B) Generally, for commercial power system, the terminal voltage (phase to phase) adopted for of TGs may be 11 KVs; but for larger machines, the generation terminal voltage adopted can vary from 16 KVs to 27 KVs.*

* **NOTES:**

   (*i*)   There are sets operating with terminal voltage of around 33 KVs, with special grade insulation system adopted for stator windings to withstand higher dielectric stress; this is a costly option, but the cost can be off-set against cost savings resulting from reduction in the size of the TG specially by way of down sizing it's electro-magnetic circuit structure and the stator winding conductors.

   (*ii*) With higher dielectric insulation strength facilitating adoption of higher generation voltage, the size of the generator can be reduced appreciably, since higher degree of temperature rise can be allowed enabling higher magnitude of current flow in the winding conductors. But on the other side, higher costs will be involved in adopting the matching higher voltage rating of the Generator Transformer and associated circuit-breakers and switch-gears.

(C) **TG Stator Comprise of:**

   (*i*)   A fabricated steel body made of steel plates shaped in to a cylindrical structure inside which ribs are welded to accommodate and support ETS stampings to build up the magnetic structure. In the slots provided in the stacked up ETS stampings, insulated windings are laid with properly configured insulating fixtures and bandaging threads (made of Teflon/Dacron coated with insulating varnish of appropriate grade).

(ETS = Electro – Technical Steel)

(*ii*)   Stator windings bars are laid in a double star waive layout. Windings are made of bunches of rectangular cross-section, or bunches of a combination of square and rectangular copper conductors, duly insulated (impregnated under vacuum suction pressure) by special grade epoxy compound; such conductor bars, after due curing, are ready to be laid in the stator stamping slots.

(*iii*)   Generator end-shields mounted bearings that supports the TG Rotor (with associated lubrication and cooling system)

(*iv*)   Gas cooler assemblies are mounted on the stator frame. Also, if the stator windings are chilled water cooled, (in addition to chilled Hydrogen gas cooling), then such water cooling devices are also mounted on to the stator body, normally from the bottom side of the stator body.

(D) **TG Stator Electrical Connection System:** Terminals of the laid out windings are brought out through properly shaped ceramic/epoxy bushings of appropriate voltage ratings; then led inside a Bus-duct system with support insulators , to the Generator Transformer though the dedicated Circuit-Breaker and associated switch-gears and protection equipment; and then eventually, the TG gets connected to the station bus-system, which in turn gets connected to the grid (again through the station switch-yards/circuit-breakers) for onward transmission of the generated power to the load centres.

(E) **TG Rotor System**

(*i*) As we have mentioned earlier, the TG rotor has to function like a rotating magnet with it's insulated windings being excited by Direct Current supply,

(*ii*) Insulated rotor windings are laid in lap coil system, (*iii*) **Excitation System:** Present technological trend is either to deploy static excitation system externally, or, mount heavy duty Diodes on the rotor for brush-less injection of DC supply to the rotor windings. (Diodes convert AC field generated in the rotor to DC field,) thus exciting the rotor to act as a rotating magnet (which induces AC EMF in the stator windings as already mentioned earlier).

(*iv*) Rotor shaft of the TG is coupled to the LP rotor shaft of the turbine and gets the rotational torque from the same.

(*v*) TG rotor is mounted on both sides by bearing pedestals; bearings are lubricated continuously and cooled by water circulation in the water-tight bearing housing.

(F) **TG Electrical Insulation System:** Generally, both stator and rotor windings are recommended for Insulation **Class-F (Allowed Temperature Rise (Above Ambient 30°C) Upto 155°C.**

(G) **TG Local Control Panels** are installed in the turbine hall near the TG ; these are inter-connected to the control panel system installed in the **Central Control Room.**

3.6 STATION AUXILIARIES AND SERVICES

In addition to the details mentioned at paragraph-3.2.2 (4) above:

(*i*) **Back up Power Supply/UPS System:** It has to be ensured that Battery-Bank is healthy and live all the time; during station power failure and the generator tripping creating no power situation, lubrication oil supply to the bearings must continue till the TG rotor comes to a stand still condition (on tripping at full speed, it may require around 30 minutes to come to rest); same is the case with ST rotors also; during the slowing down period, the emergency lubricating pumps operated by DC motors getting supply from the UPS, have to come into operation without fail; failure may cause starving of lubricants in the bearings and thus cause heavy damage to the bearings and also to the rotor shafts; repairs will be very costly affair, besides leading to prolonged shut down (loss of power generation).

(*ii*) Coal yard/Fuel supply system needs regular feeding and upkeep for dependable service.#

#REMARKS: To give some idea about the quantum of coal consumption and the corresponding steaming capacity (from the dedicated boiler) required for coal fired Sub-Critical STG Sets at Full Load, the following table seeks to present some data: (Assuming the Calorific Value of the coal at 3,000 Killo-Cal/Kg):

Parameters	110 MWs	210 MWS	250 MWS	500 MWs
(a) Coal consumption (Tonns/hr)	88	168	200	400
(b) Steaming capacity (Tonns/hr)	375	700	805	1,725

NOTE: The above data may vary depending on the design/configuration and technology adopted for the Boiler and the Steam Turbine; also on the quality of coal used.

REMARK: As soon as the TG trips and goes out of the system, the emergency stop valves and other inlet valves of the ST get activated stopping the steam supply to the ST Cylinders and eventually putting the ST out of Operation and eventually stopping ST through the Barring gear system getting locked with the rotor shaft.

(*iii*) Station Switch yard and Station power supply system and associated equipment have to get serious attention regularly for ensuring dependable service.

(*iv*) Security and Safety (including Fire Safety) must get proper attention of the Station Management.

(*v*) Communication System has to get due importance and dependability ensured.

3.7 ECOLOGICAL/ENVIRONMENTAL ISSUES RELATED TO STG PLANTS

This has emerged as a major issue before the various National and International Forums deliberating on Global Warming and Climate Change trends. With respect to STG Plants, the following issues need attention and satisfactory Resolution/corrective mitigation:

(A) In the case of Fossil fuel (particularly coal, lignite) based plants:

(*i*) Level of atmospheric/air pollution due to emissions of CO_2, CO, NO_X, SO_X, particulate matters, waste heat discharge have to be estimated at the DPR stage and due clearances have to be obtained from the Central/State Pollution Control Authorities.

(*ii*)   The same process has also to be carried out for ash and slag disposal (which is likely to have adverse affect on the land and populated areas in the vicinity).

*(*iii*)   Coal storage yard can cause environmental pollution in the vicinity areas particularly during rainy seasons. Adequate pollution control measures by way of installation of equipment/ structures, controlled/quarantined dumping pits and grounds, have to be taken to meet the prescribed standards.

(B)   **In case of Natural Gas and Furnace Oil Based Plants:** Issues by and large, are the same as above under paras (A-i, ii), requiring adequate measures for containment/emissions control within prescribed standards.

*NOTES:

(*i*)   Installation of Electro-Static Precipitators (ESPs) and Gas Scrubbers (GS) have to be considered;

(*ii*)   Height of the Chimneys are also important factor to considered for wide dispersal of residual emissions emanating from the ESPs and GSs installed.

(C)   Like the main plant and machinery, pollution control installation (ESPs, GSs, quarantined dumping pits/grounds) also need regular attention, maintenance/upkeep and servicing ("Not the put and Forget" Attitude)

3.8 MAINTENANCE, SERVICING AND ADMINISTRATIVE ISSUES

(A)   It is to be appreciated that a STG Unit is a high temperature, high pressure, high speed (normally 1500 rpm/1800 rpm to 3000 rpm/3600 rpm depending on the design and system frequency desired) rotating machine, designed for continuous duty at full load capacity. Therefore, occasional servicing, repair, preventive maintenance, replacement of quick wearing/ worn out parts/auxiliary devices, as recommended by the manufacturer, to be followed with due seriousness by the operating engineers and technicians.

(B)   Maintenance engineers and technicians are to be trained for the job and they should be adequately conversant with trouble-shooting procedures.

(C)   Periodical shut-downs are to be planned for check up and preventive maintenance.

(D) Stocking of spare parts for routine and break-down maintenance must be ensured including those for auxiliary equipment BUT keeping an eye on the prescribed Inventory Control Regimes.

(E) For rotating equipment/parts, adequate attention must be paid for trouble-free/satisfactory working of:

   (i)   Main Oil Pump (MOP), Auxiliary Oil Pump (AOP), Jacking Oil Pump (JOP).

   (ii) Bearings.

   (iii) Dynamic Balancing as and when felt necessary/as recommended by the manufacturers (Specially During Major Overhauling).

   (iv) Indicating Instrumentation (Temperature and Pressure, Current, voltage, frequency).

   NOTE: Digital ones with back up systems are preferred; if feasible change over to Digital System and replacing Analogue system, (which is less accurate) may be taken up.

(F) For major overhauling or for major repairs, it is always prudent to associate the experts of the manufacturers, for best results.

(G) Operating engineers and technicians should be strictly advised to refrain from tinkering with or digress from the prescribed/recommended operating procedures to avoid unexpected mishaps (there are ample examples of such wrong-doing-both in India and Abroad); over-zealousness/ over-confidence on the part of operating staff can be suicidal.

(H) In case of gas and oil storage yards (tanks, pipelines, etc), extra care has to be taken to avoid chances of accidental fire incidence; leakage and spills are to be avoided/ contained within the prescribed safety standards; No slackening on this aspect.

(I) Safety and Security for the whole plant:

   (a)   Adequate measures should be in place;

   (b)   Operating staff should be exposed to training programmes, appreciation seminars and drills/periodically: at least once in six months.

(*c*) Adequate numbers of Warning/Advisory Signages be put up at vantage points (with Telephone Nos. of key Officials). Station Superintendent, Safety officer and Security Officer should invariably have.

  (*i*) Map of the entire plant area with critical installations marked on it.

  (*ii*) A display board indicating telephone Nos. of key officials, civil and police authorities, hospitals, ambulance services.

(J) Dumping of scraps/wastes etc. should be far away from the plant site, they can be source of fire hazards also.

(K) As regards Incentive/Reward Schemes, for God's sake, DO NOT IGNORE the Maintenance staff, that can be a big hindrance to smooth running and productivity.

(L) Management must treat the power station as a critical Service Industry, and try to maintain a good level of Industrial Relation.

3.9 OTHER RELEVANT ISSUES

(A) STG plants are a necessary evil from the point of view of environmental aspects, keeping in view that major capacity of generation comes from this method (in India, around 70% of total installed capacity comes from STG plans).

(B) Super Thermal Power Stations (predominantly STG based) in India have installed capacity over 2,000 MWs to 4,000 MWs at a single location, comprising mostly of 250 MW, 500 MWs, 600/660 MWs, 700 MWs, 800 MWs units*. However this standard may change with more and more higher capacity units are installed in one STG plant.

* NOTES:

  (*i*) Since it poses practical operational difficulties particularly for trouble-free load sharing in parallel operations, it is not advisable to have units of widely differing in capacity ratings installed side by side in the same STG plant, to be connected to the same Station Bus Bar System; IT MAY CREATE SYSTEM IMBALANCE during load sharing and fault condition developing in individual unit bus bars or in the station common bus system.

(*ii*)   In case it is decided to have more than one STG Stations located at the same place (side by side), each of such station should preferably have same or near same rating units; all the Station Bus Bars to get connected to the Station Switch Yard Bus System through appropriate station transformers, matching circuit breakers and switch gears of appropriate SC (Short Circuit) Ratings. However, before such arrangements/layouts are decided, experts group has to carry out analysis of expected maximum fault current/fault KVA/MVA levels under different loading and fault conditions and then decide on the switch gear/protection gear sizing/ratings.

(*iii*)   Readers (electrical engineers in particular) are advised to get reasonably familiarized with the Technology of Transmission and Distribution and the standard procedures for deciding the station/sub-station layouts. **This book does not propose to go into such topics which itself relate to a separate field of Electrical Engineering**.

(C)   Modern STG Plants employ computerized automated operating and control systems with digital instrumentation and display panels for performance measurement and indications, giving a delightful work experience to those engineers sitting (and observing) in the Central Control Room, giving them also an exalted feeling that they are controlling a mammoth working contraption mostly by their finger tips.

(D)   The Scientists, the Engineers and the Technologists have conceived and eventually put into practical shape such a marvelous operating system to serve us days and nights, with rare occasions of failures, which we take in our strides.

(E)   Those young engineers/engineering students who aspire to make a successful (professional) career in the field of STG plant Operations and Management, are advised to get exposed to relevant training programmes; **Power Training institutes** of Government of India, located in different regions, are offering such training programmes.

□□□

‖
4
‖

Gas Turbine Driven Thermal Power Plants (GTG Plants) Combined Cycle Power Plants (CCPP) Diesel Generating Sets (DG Sets)

"Science alone is not Technology and

Technology alone is not Innovation."

— *Akio Morita, Former Chairman, Sony Corporation.*

4.1 GAS TURBINE DRIVEN THERMAL POWER GENERATING PLANTS (GTG PLANTS)

4.1.1 Gas Turbine (GT) driven Electrical Power Plants have emerged, in the last five decades or so, as a more convenient and compact system of thermal power generation.

4.1.2. General Configuration and Layout of a GTG Plant

(A) Gas turbine is a special type of high speed rotating engine to serve as prime-mover for the coupled electrical generator. Gas turbine utilizes the compressed air raised to around 300°C to 350°C; this compressed air is generated in the compressor portion of the turbine. On the other side, selected fuel (see paragraph-C below) in gaseous state is injected in a combustion chamber where the heated compressed air gets mixed with the surcharged preheated fuel gas creating an explosive mixture which gets ignited by means of special igniters electrically (utilizing highly combustible propane gas and high voltage sparking system located inside the combustion chamber, some what like an Internal Combustion Engine). Combustion takes place inside the combustion chamber and the resultant hot flue gas at temperature at around 1050°C/1100°C is passed at high pressure (resulting from explosion inside the combustion chamber) into the turbine stage having 3-4 stages each of moving and fixed blades made of high grade alloy steel material (to withstand such high temperature and resultant thermal stress). The thermodynamic force thus created acts on the turbine blades (both fixed and moving) resulting in generation of the required torque on the turbine shaft to deliver the required rotating mechanical force to the coupled rotor of the electrical generator. Normally, the generator is located on the opposite side of the turbine (cold end).

(B)   The gas turbine has three major operating components on it:

  (i)   The air compressor with multiple stages (12-16 stages) of moving and fixed blades.

  (ii)   The turbine with three to four stages each of moving and fixed blades is placed between the last stage of the compressor and the exhaust chute.

  (iii) Combustion chambers (normally 2 nos., one from each side).

- **Figure No. 4.1 represents a Schematic Layout of a typical Open Cycle GTG Power Plant; Figure No. 4.l(A) shows a Sectional view of a Typical Gas Turbine.**

(C)   The technological innovation done by some manufactures (GE, SIEMENS, Westinghouse, Rolls Royce, -to mention a few) is to have a ring type combustion chamber with multiple nozzle burners placed around the turbine stages at a place between the last sage of air compressor and the entry stages of the turbine. This system reduces the size of the turbine auxiliaries and eliminates the use of externally place combustion chambers, thus make the GT a more compact machine with higher thermal efficiency.

NOTE: Aircraft gas turbine engines (aero-derivative gas turbine commonly known as Jet Engines) are a different class by themselves, although they are also basically power plants; we are excluding them from the scope of the present book since they are not normally being used for conventional utility electrical power generation plants. The large aircraft gas turbine engines presently being deployed, can generate mechanical power (torque) equivalent around 30 to 50 MWs.

(D) The gas turbine, depending on the design adopted, can be operated by using Diesel oil, Natural gas/Naptha, LPG, LSHS Diesel, which make the gas turbine plants attractive from the point of view of low level of pollution and clean operation (quite low level of NOx, SOx and CO/CO_2 emissions can be achieved). Gas turbine plants are compact requiring much less covered area and has much less numbers of auxiliary systems as compared to the **conventional steam turbine stations**. Gas turbine set has as distinct advantage of quick start/ restart within 10 to 15 minutes as compared 6 to 8 hours (cold start) in the case of conventional STG thermal sets, thus making it specially suitable for meeting peak load demand. The station power requirement is also much less, 6 to 7% as compare to 10-12% in case of steam turbine power station.

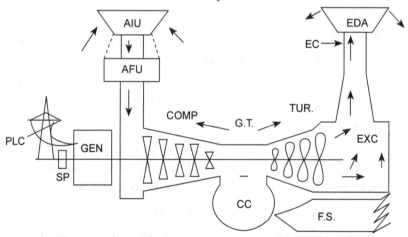

Fig. 4.1: Schematic Arrangement of a Gas Turbine Generator Plant (Open Cycle)

LEGEND:

GT : Gas Turbine comprising of COMP (Compressor Stages) and TUR (Turbine Stages); CC - Combustion chamber;

EXC : Exhaust chamber; FS- Fuel supply system; EC - Exhaust chute;

EDA : Exhaust discharge to atmosphere;

AFU : Air filtration unit; AIU - Air intake unit; GEN - Generator;

SP : Starting prime-mover for GT (diesel engine); PLC - Power load center.

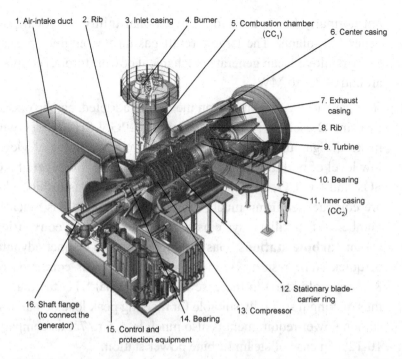

1. Air-intake duct   2. Rib   3. Inlet casing   4. Burner   5. Combustion chamber (CC₁)   6. Center casing   7. Exhaust casing   8. Rib   9. Turbine   10. Bearing   11. Inner casing (CC₂)   12. Stationary blade-carrier ring   13. Compressor   14. Bearing   15. Control and protection equipment   16. Shaft flange (to connect the generator)

Fig. 4.1: A Schematic Arrangement of a Gas Turbine Generator Plant (Open Cycle)

(E) The Electrical Generator, normally a 3-phase AC Synchronous machine, is either coupled directly to the shaft of gas turbine, or coupled through a speed reduction gear+ train system since the speed of a gas turbine can be much higher than 3000/3600 RPM required for 50/60 Hz operation. Other sub-systems and auxiliaries like excitation, cooling/ventilation, control panels, switch-gears/ protection system etc. are similar to those adopted for the turbo-generator described for STG sets (refer to para 3.5 above).

NOTE: To avoid the speed reduction gear-train system (which brings in additional operational and maintenance complexities and costs), the Gas Turbine itself is so designed that it can be operated at the desired speed through incorporation of an appropriately designed and configured speed control mechanism (fuel injection and control).

(F) The above mentioned Gas Turbine driven power generation is known as '**OPEN-CYCLE**' operating system as in this case, the exhaust from the gas turbine is directly discharged (with due emission control) to the atmosphere along with the residual heat in the flue gas. In this system, sufficient residual heat (exhaust temperature can be above 500°C) is lost. In the Combined Cycle Power Plant (CCPP), this residual heat is utilized to run a smaller steam turbine generator set (around 33% to 40% additional power to the system), **(more details about CCPP are stated in paragraph 4.2 below)**.

(G) A Gas Turbine requires a cranking prime-mover for its start up (that is, to give it the initial rotating motion for firing up like an 1C Engine); this prime-mover is generally a small diesel engine of capacity of around 1.5% to 2% of GT rated Capacity (in terms of KWs/MWs). However, as per the latest trend, if station power is available, the coupled electrical generator itself is so designed and constructed so as to function as a starting motor to give the initial rotation, and once the GT starts running by itself, motoring action is stopped, and the generator is then operated as a generator to generate power in the usual manner.

Fig. 4.4: Photograph of a bladed rotor of a gas Turbine

(H) The degree of reliability for trouble free and satisfactory operation even at full load is very high in comparison to the conventional STG thermal sets. Gas turbine - Generator sets can safely operate at + 3% to −3% of rated speed, without losing stability (synchronism) in the system.

4.1.3 Certain Special Features of GTG Plants

(*i*) GT is basically an 1C (Internal Combustion) Engine operating at a much higher temperature at the turbine stages (around 1100°C); hence cost of turbine stage construction materials (special grade alloy steels) are much higher as compared those for Conventional sub-critical ST.

(*ii*) Fuel cost is higher (diesel, Refined Natural Gas, aviation turbine fuel), depending on the GT design and technology adopted.

(*iii*) In case of GTG plants, the number of auxiliaries are much less than those required for STG plants, hence land area and covered space required will be much less (and hence lesser costs towards land and buildings).

(*iv*) Plant life cycle will be lower, mainly because of high temperature operation of GT at continuous duty cycle; hence the depreciation rates to be charged on the Gross Block (capitalized value of the plant, specially of the GT entered into the Asset Ledger) will be higher.

(*v*) Operating and maintenance costs are normally lower for similar capacity, as compared to the STG sets.

(*vi*) There are technological constraints in adopting GTG sets of unit capacity above 150 to 160 MWs; normally NOT recommended; preferred range is 100 to 125 MW unit capacity.

(*vii*) GTG can be a economic and advantageous option when operated in the Combined Cycle Power Plants (CCPP): a GT-ST Combination; (Pleas see para 4.2, below for more details on (CCPP).

(*viii*) GTG plants are normally preferred for medium capacity generating stations of gross installed capacity not exceeding 750 MWs at one station (i.e. around 6 × 125 MW). Therefore, GTG- plants are not likely to be a part of a STG-based Super Thermal Power Station (which are normally with the aggregate capacity above 2000 MWs in one station).

(*ix*) GTG stations are preferred as a peak-load supplementary power stations; also normally not preferred as a base load station, to avoid continuous duty-cycle operation (which reduces plant operating life cycle and increase cost of generation and maintenance).

(*x*) GTG plants require much less time for start up as well as for taking to the Full load conditions.

(*xi*) GTG plants are less polluting, and hence preferred for urban areas provided assured fuel (commonly, Natural Gas, CNG) supply is available in cost-effective regimen.

(*xii*) There are a number of manufacturers of GTs (for power generation) in the world for different unit capacity ranges starting from 5 MWs to 160 MWs; however most reputable ones are General Electric, Siemens AG, Rolls Royce, Pratt Whitney, Mitsubishi, Hitachi, Westinghouse, English Electric; BHEL (India) is presently (2017) manufacturing GTGs to General Electric Design and Technology (up to around 125 MW unit capacity) at its plant at Hyderabad. BHEL Hardwar plant had, some years ago, manufactured several 154 MW capacity GTG sets to Siemens AG design and technology (mostly exported), but discontinued because of lack of demands for such high capacity GTG sets.

4.2 COMBINED CYCLE POWER PLANTS (CCPP/IGCCPP)

(A) These are a special class of power plants with higher thermal and overall energy efficiency for the same range of power output in comparison to a conventional thermal power station, besides being quite reliable for long hours of operation, and causing less pollution. Therefore, there is a greater preferences for this type of power station all over the world at present since the system utilizes the residual heat from the GT exhaust (which will otherwise go waste being discharged to the atmosphere).

(B) The Combined Cycle Power Plants comprise of smaller capacity one or more gas turbine generator sets combined with a steam turbine generator set, - both set-up running together in harmony to generate electrical power from separate electrical generators, one coupled to the gas turbine and other coupled to the steam turbine, run in parallel and in synchronism to one another and are connected to the same station bus system/grid system. Hence, this operating system is called the COMBINED CYCLE POWER PLANT (CCCP).

The Schematic Layout of a CCPP is presented at Figure No. 4.2.(also refer to paragraph (E) below).

(C) The exhaust gas emanating from the gas turbine have a temperature range of the order of 500-540°C (depending on design of the turbine and type of fuel used). This flue gas is passed through a

Heat Recovery Steam Generator (HRSG - which is a sort of a heat exchanger) where water flowing through the metallic tubes (stainless alloy steel, titanium alloy, Cupro-Nickel alloy, depending on the system requirements) placed inside the HRSG, gets transformed into steam at around 350°C to 535°C superheat at around 33 ATA to 45 ATA depending on as to whether the HRSG is an 'Unfired' or 'Fired' system. This steam is fed to a steam turbine which run at the designed speed (3000/3600 RPM or 1500/1800 RPM depending on design and the system frequency adopted) to rotate a matching electrical generator to generate electrical power. As already stated above, this generator can run in parallel and in synchronism with the generator directly couple to the gas turbine. Since no boiler with the usual fuel supply paraphernalia is required to generate steam in the conventional way, this system of power station is cleaner, less polluting, require much less covered area and lesser numbers of auxiliaries, lesser operating cost, quicker start-restart facility, besides operating at higher thermal efficiency.

(D) Although the CCPP is comparatively more efficient thermal power generating system, but because of certain technological and operational constraints, the same cannot be adopted yet for a large power station; so far, an aggregate capacity of a combined cycle power plant (CCPP) anywhere is not more than 750 MWs (GT unit capacity above 150-160 MW normally not available, nor preferred).

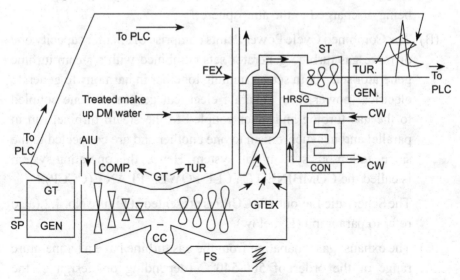

Fig. 4.2: Schematic Arrangement of a Combined Cycle Power Plant.

Legend:

GT : Gas Turbine comprising of COMP (Compressor Stage) and TUR (Turbine Stage); ST – Steam Turbine; CON – Condenser; GTGEN – Gas Turbine coupled generator; TURGEN – Steam Turbine coupled generator; HRSG – Heat Recovery Steam Generator; CW – Cooling Water;

GTEX : Gas Turbine exhaust; CC – Combustion Chamber; FEX – Flue Gas exhaust (to atmosphere); SP – Starting prime-mover (diesel engine); AIU – Air intake unit; FS – Fuel supply system; PLC – Power load centers.

(E) Considering the overall system optimization, the preferred combination for a CCP Plant is as under:

(*i*) Two identical GTs each coupled with a matching identical generators running in parallel;

(*ii*) One smaller STG set getting super-heated steam from the common $HRSG^+$ which gets the heat source from the GT exhaust flue gases being discharged from the GTs, eventually into the HRSG.

(*iii*) All the three generators in the combined system operate in parallel (in synchronism) and are connected to the common station bus-system, and in turn, to the proximity Grid System.

• **A Commonly Adopted Schematic Layout with 2+1 Combination for CCPP Is Presented at Figure No. 4.2(A).**

4.2.1 Integrated Gasification Combined Cycle Power Plants (IGCCPP)

This a special category of CC Plant which aims at optimizing the gas-turbine combined-cycle technology with the adoption of various gasification system based on gasification of solid fuels(coal/lignite/pit) and liquid fossil fuels.

This technology has the potential of effecting as much as 25% savings in overall fuel consumption in the plant capacity of 200 MWs and above, with possibility of reaching the overall plant operating efficiency of around 55%.

Although at present only a limited number of such plants are in operation, preference for this technology will grow in the coming decades specially with the development and use of **Pressurized Fluidized Bed Gasification (PFBG)** process, since this technology also helps in substantial reduction in the level of polluting emissions.

>> Figure No. 4.2(B) Depicts a Simplified Schematic Arrangement of a IGCC Power Plant.

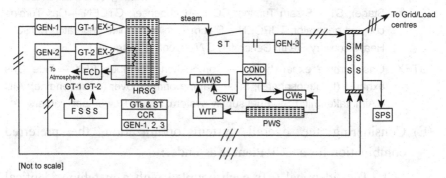

[Not to scale]

Fig. 4.2 (A): A commonly Adopted Schematic Layout with 2+1 Combination for CCPP is Presented

LEGEND:

GT =	Gas Turbine ; GEN = Generator; ST = Steam Turbine; EX = Exhaust (flue gas} Chutes.
HRSG =	Heat Recovery Steam Generator; DMWS = Demineralized Water Supply System; WTP = Water Treatment Plant.
WTP =	Water Treatment Plant; CSW = Condensate Water; CWS = Cooling Water Supply System; PWS = Primary Water Supply Source; COND = Condenser; SBS = Station Bus System.
MSS =	Main Sub-Station; SPS = Station Power Supply Sub-Station.
SBS =	Station Bus System.
FSSS =	Fuel Storage and Supply System; CCR = Central Control Room ; ECD = Emission Control Device.

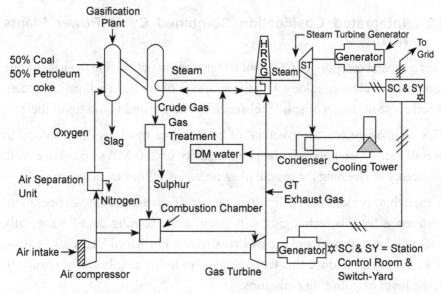

Fig. 4.2 (B): Simplified Process Flow Diagram of a IGCC Power Plant

4.3 DIESEL ENGINE DRIVEN POWER PLANTS

(A) Diesel Engine driven power generator sets are available and are in use in the unit range of 5 KWs to 6000 KWs+ (higher sizes are also adopted in case sufficient quantity of diesel oil at a viable cost can be made available). Since diesel oil is generally costly, DG sets are predominantly used:

- As standby sets for emergency power generation.
- To meet peaking load demand in certain locations,
- To meet essential requirement of power in isolated/remote load-centres where grid connectivity is not available.

+ **NOTE:** Because of technological and operational constraints, single DG engine above 6000 KWs is not normally recommended.

(B) In addition to various points mentioned at paragraph D-1 in **Chapter 2**:

- Sets up to 150 KWs can be in portable version.
- Higher capacity sets beyond 15 KW, the diesel engine may have multiple cylinders in even numbers, configured both in single line vertical or duel line V-formation, rotating the same crank shaft. A 2000 KW engine can have up to 16 cylinders in V-formation.

  A number of DG in sets can be installed in one power house for parallel operation.
- DG sets are uneconomical for large scale power generation except in those countries where abundant supply of diesel oil at low cost is available (Saudi Arabia, Kuwait for example).
- As compared to steam and gas turbines, diesel engines are more prone to frequent breakdowns and operating trouble resulting in higher down time, particularly, when sets are run continuously for long hours.
- Diesel Generator set can be started or restarted, as well as can be taken to full load capacity in a few minutes. In fact, there are sets in the range of 100 to 200 KWs for emergency power supply system in hospitals, airports, hotels, mines, and special purpose scientific research stations which, in case of

power supply failure from the grid/distribution systems, can start automatically within 15 to 30 seconds and can be loaded to full capacity within a few minutes.

- Such sets will have backup UPS battery bank system to supply starting cranking power.

- Diesel engines will have water cooling system in the engine cylinder jackets with necessary pumps, water source and associated pipelines.

- Diesel engines have self-driven++/ externally driven lubrication system to supply lube oil to the engine bearings and to the other moving parts where continuous lubrication is required.

- Diesel engines up to 5 KW or so can be started by hand cranking also.

   NOTE: Through appropriate kinematic linkage mechanisms.

- Diesel engines of the capacity range above 15 KWs and up to 2000 KWs, will require either a pony motor which may be a small petrol/ diesel engine, or an electrical motor for starting; this pony motor gets coupled to the engine crank shaft for a few minutes through a dis-engageable clutch/ gear box system. (NOTE: electric motor used may have to get power from a battery bank in a no-power situation).

- The diesel engines upto 150 KWs (multi-cylinders) can also be started by a small hydraulic turbine system using hand primed up hydraulic cylinder that sends hydraulic fluid jet at high pressure to the hydraulic turbine which gets engaged to the crank shaft of the diesel engine giving it rotation for a few seconds, and that causes the diesel engine to start up, provided that other pre-requisites for a successful start up are lined up properly.

- Diesel engine above 2000 KWs having multiple cylinders will have to adopt compressed air system of starting: compressed air stored in air-bottles at 40-50 kg/cm^2 which can supply sufficient rotating torque to the engine crank shaft to give it a few rotations before the engine gets fired up and starts running eventually. The compressed air is released by operating the starting lever manually, or by a automated push-button system; before this is attempted, it has to be ensured that the air bottles are filled with compressed air at the requisite pressure level. Also, before operating the air release lever, it has to be ensured that the engine fly wheel is adjusted to the prescribed firing order position (very important condition for a successful start).

(C) Diesel Engine for power generation is normally designed for a speed range of around 428.6 RPM to 1500/1800 RPM (higher the capacity/size, lower is the RPM). Therefore, generator coupled to it is generally a salient-pole 3-phase AC synchronous machine for 50 Hz or 60 Hz system. The generator has to have the system of matching excitation system controls, switch gears and protection system, cooling/ventilation system etc. However, depending on the size and capacity of the generator, cooling and ventilation can be by natural air forced into the machine by fans coupled by suitable V-belt -pulley system to the generator shaft. Lubrication of bearings is done by pumping lube oil under prescribed pressure, and cooled by passing cold water through jackets around the inner housing of the bearings. However, for smaller generators say below 300 KWs, such oil lubrication and water cooling in the generator bearings may not be required and may be avoided to keep the cost lower; in such a case simple journal ball or roller bearings with proper grade of externally injected (time to time) greasing may be enough.

(D) **Figure No 4.3 Represents a Schematic Layout for a Standard DG set.**

(E) DG sets up to 200 KWs capacity are available for silent mode of operation (noise level controlled below 60 decibels).

(F) **Use of Turbo-Charger for Air Intake**

In case of large Diesel Engines, a turbo-charger using a part of the hot exhaust flue gas to pre-heat the air injection into the engine cylinders is mounted on the engine; this improves the overall fuel efficiency and engine performance.

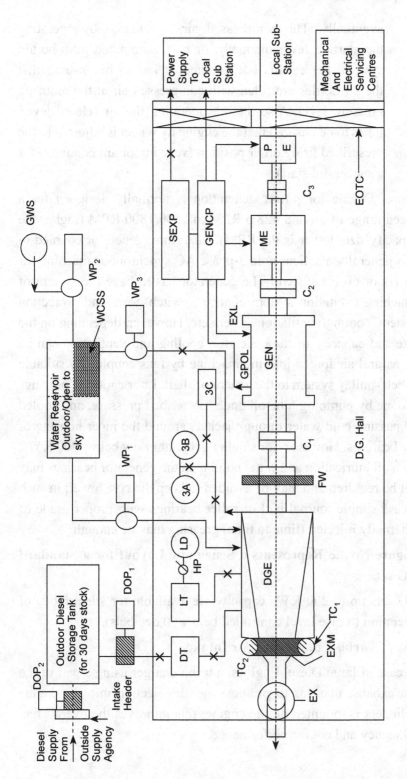

Fig. 4.3: Schematic Layout of a Diesel Engine Driven Electrical Power Generating Station
(Above 500 kw Unit Capacity, – with Compressed Air Starting System)

LEGENDS:

DGE = Diesel Engine (Muiti-cylinder); DT = Diesel oil Day tank; GEN = Generator; SEXP = Static Exciter Panel (alternative exciter system); ME,= Main Exciter; C_1, C_2, C_3 = Couplings; EX = Exhaust system with silencer; DOP_1, DOP_2 = Diesel oil pumps; GENCP = Generator Control Panel; WP_1, WP_2, WP_3, - Water Pumps; LT = Lube oil tank; EXM = Exhaust Manifold; LD = Lube oil Drum; 3A,3B = Compressed air storage cylinders; 3C = Compressed air pump (driven by a small petrol or diesel engine); WCSS = Water cooling sprinkler system; GWS " Ground water supply system (submersible pumping system); GPOL = Generated power output lines(TPN);

EXL = Excitation power supply lines: TC_1, TC_2 = Turbo-chargers (for pre-heating of air intakes); EOTC = Electric overhead travelinger-crane LD = Luv oil drum: HP = Hand pump; FW = Fly Wheel: PE = Pilot exiter.

4.4 ECOLOGICAL ISSUES THAT NEEDS ATTENTION FOR REMEDIAL/CONTROLLING MEASURES FOR THE ABOVE THREE SYSTEMS OF POWER GENERATION METHODS

(*i*) **For Open Cycle GT sets the exhaust flue gases contain:**

- Residual heat being discharged into the atmosphere.

- Emission of CO_2 Co, NO_X, SO_X have adverse impacts on environment since these sets use fossil fuels.

- Emissions of particulate matter (mostly fine ash and carbon particles) have ecological impacts on surrounding areas.

- Uncared for oil (fuel, lubricating) spills may create contamination in the vicinity land areas.

(*ii*) **In case of CCPP stations**

- Pollution level (air, land) is much lower but may not be always within permissible limits; appropriate controlling measures have to be put in place to conform to permissible limits.

- Ash and slag disposal poses problem as usual.

- Here also, uncared for oil spills can contaminate vicinity land areas.

(*iii*) **In case of DG sets**

- Emissions of CO_2, CO, NO_X, SO_X in the exhaust will have ecological impacts, diesel as a fuel causes emissions of carbon particulate matter at a much higher level.

- DG sets are quite noisy, causes high level of noise pollution, needs mufflers in the exhaust charts);

- Here also, uncared for oil spills can contaminate surrounding areas.

❑❑❑

5

Hydro-Power Generation Plants

> *"Flow of innovative urges leads to flow of Energy that drives the human endeavours."*
>
> — *From a lecture on Innovation delivered by an Industrial Engineering Professor.*

5.1 HYDRO-POWER GENERATING SYSTEM

(A) The system uses the Kinetic Energy of the falling waters (mgh, the potential energy gets converted into Kinetic energy during the fall). The falling waters transfer the Kinetic energy to the blades of the runner of the hydraulic turbine causing the runner to rotate at a speed (rpm) depending on the head, discharge rate and the geometric profiles of the runner/runner blades. The runner being firmly mounted on the turbine shaft, which in turn transfers the rotating mechanical energy/force (in the form of Torque) to the coupled rotor shaft of the Hydro Generator (HG), normally a 3-phase AC Synchronous machine. The rotor windings of the hydro-generator being excited with DC supply, makes the rotor a rotating magnet. Following the Faraday's law of Electromagnetic Induction, the rotating electro-magnetic field of the HG rotor induces EMF in the stator windings of the hydro generator, creating potential difference in the terminals (voltage) thus making it ready to deliver Electrical Power to the connected Loads through the Generator Transformer, associated switchgears/circuit breakers, other controlling and protection devices. The generated power then is evacuated through the connected Grid Networks (transmission/distribution lines which can be over-head networks, or for short distance, underground cables) to the targeted load-centres.

(B) Hydro Generator (HG) by design and construction is a robust electrical machine which can generate hundreds of MWs of power in continuous duty cycle, provided enough water discharge rate can be arranged and ensured for the Hydro Turbine during the generation cycle.

5.2 HYDRO-POWER PLANTS

Hydro-Power is one of the most Cost-Effective system of electrical power generation, depending on the geographical location of the hydro-energy source (water falls, rivers, streams) with adequate Head and Discharge rate available. The flow conditions of the water source can also be suitably modified and utilized by artificially creating the required heads and discharge rate in consideration of the potential of the source.

In order to harness the energy of the fall and flow of water to the optimum level, suitably designed and constructed turbine and coupled matching generator, are placed at a suitable location in the dam/barrage system; the system can generate enormous amounts of electrical power (in some cases, hundreds of MWs; (sets with unit capacity of 850 MWs already operating in the USA).

>> **Hydro-Turbines can be designed and constructed as per the following predominantly adopted configurations depending on the prevailing hydrological and geographical conditions at the project sites :**

5.2.1 Kaplan Turbine Based Units

(A) Kaplan Turbine is used mostly in vertical shaft arrangement in hydro stations where the Head is rather low (not more than 50-60 metres). In case the discharge is appreciable, multiple units are installed in the same power house. There can be more than one power station under the same dam/barrage system. The generator is mounted above the turbine in vertical shaft position and the generator shaft is coupled to that of the turbine below. The exciter is mounted, again in vertical shaft position, and is coupled to the generator shaft. When you visit a Hydro-power station, you first enter the exciter hall-cum-control room located, normally at the upper most working floor. However, there can be another floor above the exciter hall on which Gantry Cranes, and Electrical Transformers with circuit breakers and other matching Sub-Station equipment can be installed for better and economic usage/saving of space, lessening construction cost, facilitating operational convenience.

(B) The Kaplan turbine is equipped with Electro-hydraulic speed and torque control governing system which continually adjusts the turbine operation (to maintain constant rpm) depending on load demands and the matching discharge rate required. Although the generator has its own control equipment, but both the turbine and generator control systems have to work in perfect coordination and unison/synchronization to avoid unbalanced operating conditions. Unbalanced operating condition may lead to generator losing synchronization vis-a-vis the Grid system to which it is connected and eventually trips out by the operation of sensitive protection gears (circuit breakers, relays, instrumentation, etc.). A large number of Kaplan turbine-generator sets are operating in India; to name a few: Kulhal (UK/3×10 MW), Donkarayi (AP/1×25MW), Rengali Odisha/5×50 MW), Ukai (Guj-4×75**MW), Tanakpur (UP/3×40MW).

(** highest unit capacity so far in India).

(C) Kaplan turbine is basically a reaction type rotating machine which gets its torque for rotary motion from the kinetic energy of the falling waters against the surface profiles of the blades fitted to the turbine runner. Through the electro-hydraulic governing system, the orientation of the blade faces can be adjusted automatically in response to any variation in load demands and also by adjusting the water flow by controlling the opening of the **Disc Valve** placed in the penstock before the **Spiral Casing** which directs water flow into the turbine runner. The generator is a vertically mounted 3 phase, AC, 50 Hz (in India, only 50 Hz system) synchronous machine having multiple salient poles in the rotor being coupled to the turbine runner (maximum rpm range is around 500). The excitation system normally consists of a main exciter (separately excited DC generator excited by current from a Pilot exciter which is a self excited DC generator). However as per the latest technological system, a brushless exciter with thyristorized DC excitation, or fully static excitation system for DC inputs are also being adopted. The Generator terminal (phase to phase) voltage adopted is generally around 11 KV, with the built-in percentage reactance drop of the order of 5-6% with power factor- between 0.9 to 0.8 depending on the Grid-system requirements. In order to have control on the water flow rate through the penstock, a Control Valve (commonly a Disc type one) is placed in the penstock before the turbine inlet (inlet of the Spiral Casing which directs the in-flow of water into the turbine runner).

5.2.2 Francis Turbine Based Units

(A) Francis Turbine is basically a mixed-flow Reaction Turbine mostly in vertical shaft arrangement, suitable for hydro station with medium range of heads (between 45 m to 500 m) and having substantial rate of discharge. In this turbine, the blades on the runner are fixed (integral steel casting), but water flow is controlled by a set of variable orientation guide-vanes each supported on guide vane bush-bearings fixed on the Stay-Ring, a fixed circular steel structure around the runner. These guide vanes are operated and controlled by Electro-hydraulic governing system (for controlling the flow of water into the blades similar to the one mentioned above under Kaplan Turbine). Like the Kaplan turbine, the water from the reservoir rushes to the turbine through the penstock. There can be water flow controlling gates of two types, namely: Disc and Spherical valves; one of them or both are used in the penstock depending on inlet water flow conditions : Head, Discharge rate, geographical-gradient, length of the penstock.

The runner speed of a Francis turbine can go up to 750 rpm depending on the Head and Discharge conditions; the rpm rate of the runner is critical for the design and configuration of the hydro-generator to be driven by the turbine.

(B) By construction, Francis turbine are the most robust and can handle very high discharge rates, generating hundreds of MWs of power per unit (850 MW units are operating in the Grand Couli Hydro-Power station in the USA, having 10 sets in all).

(C) The generator in this case also is a vertically mounted 3-phase AC 50/60 Hz synchronous machine, having multiple salient poles mounted on the HG rotor spider. The HG rotor is coupled with the turbine runner shaft. The excitation in this case are normally of the same category and configuration as stated above for Kaplan turbine driven HG. Except for the construction features of the turbine, all other features associated with the Francis type turbine-generator system are more or less the same or similar as in the case of Kaplan turbine based system described above.

In India, a large number of Francis-turbine-generator sets are already operating; to name a few:

Salal (J&K/6×115 MWs), Giribata (HP/2×30 MW,). Loktak (Manipur/3×35 MW), Dehar (HP/6×165 MW), Nathpa-Jhakri (HP) with 6×250 MW; Tehri Hydro project with 4 units of 250 MW each.

NOTES:

(*i*) Depending on the turbine runner speed, generator rotor diameter can vary from around 2 metres to 8 metres or even more. In case the vertical face heights of the HG poles as well as the height of vertical face of the stator magnetic circuit are increased, the diameter of HG rotor as well as that of the stator can be reduced for attaining the same level of magnetic field strength and hence for the same power out put as planned by designers; these features are decided at the design stage by the project design engineers. **This is True for All Types of Hydro-generator Units**.

(*ii*) Considering the expected water flow conditions, the designers of the HT and HG, as a standard approach, try to optimize the construction features (sizing, configurations) by using relevant quantitative techniques and applicable computer software.

5.2.3 Pelton Wheel Turbine Based Units

(A) Pelton Wheel turbine is an impulse type water turbine suitable for hydro-station with high-head (above 300 metres) and low to medium discharge. The runner speed may range between 300 to 1000 rpm. The runner of this turbine has a number of buckets with duplex impingement areas, fixed at the periphery of the runner wheel-disc. The water from the penstock is passed through a set of nozzles discharging high velocity water jets impinging on the buckets mounted on the runner wheel, thus transferring the kinetic energy of water jet to the buckets, and due to this impulsive force, runner rotates.

(B) The Pelton wheel turbine can be mounted either in horizontal, or, vertical shaft position depending on the location specific operating conditions; horizontal arrangement is preferred for low capacity units due to certain advantages and convenience. The turbine will have the usual Electro-hydraulic governing system for speed/ torque control as well as water inlet controls.

(C) The generator with its excitation system and auxiliaries have similar configurations as in the cases of Kaplan and Francis turbine operated ones. There is limited number of such sets operating in India; to name a few: Chennai (J&K /2×5 MW), Shanan (Punjab/1×50 MW) Varahi (karnataka/2×115 MW**)

(** highest capacity in India so far); there are sets with unit capacity more than 420 MW operating in other parts of the world- (3×423 MW sets with 1800 metres head working in a hydro-project in Switzerland).

5.2.4 Reversible Hydro Sets

(A) This type of hydro sets are used as a peaking power supply sets. In this type of hydro sets, the hydro-turbine (which can be both Kaplan or Francis type with some modifications in the rotor and guide vanes) can run both as a prime-mover to the generator for power generation, or, can become a water pump, pumping water back from the down stream reservoir to the up-stream reservoir during the **off peak-load period** (at night after the peak load period is over); at that time the generator operates as a motor drawing power from the bus/grid system. In India a few such sets are running: Kadana (Gujarat: 4×60 MW), Nagarjun Sagar (AP:3×100), Kadamparai (TN 4×100); besides, 3×200/220 MW sets are being planned for installation in Sardar Sarovar Dam Project (Gujarat).

(B) Such hydro unit, when operates as a pumping set, it's generator, now operating as a motor, draws around 10% more power than its rated capacity as a generator; in this case, both turbine and the generator are designed for dual performance duty cycles.

5.2.5 Bulb Type Hydro- Sets

These sets are installed completely submerged in water canal/river and the turbine-generator set is placed inside a bulb-like big water tight steel enclosure (capsule). This method is resorted to where the head is low (2 to 10 metres), but the discharge is quite high. These sets can have a unit capacity range of around 2 to 30 MWs. In this arrangement, while the turbine-generator set is placed inside the hermatically sealed steel vessel looking like a bulb, the runner of the turbine is mounted outside on the horizontal turbine shaft (somewhat like a propeller of a river craft) and remain submerged to be rotated by the kinetic energy of the flowing water rushing past the propeller like runner. The set is remote-controlled from a shore based control room; the set can also be rigidly suspended under a steel gantry like structure erected over the river/stream some what like a oil rig structure. Obviously, such set up will not be suitable for large/wide rivers/streams. These sets will be the cheapest hydro-power generation system (particularly from the structural configuration and construction point of view); and there is no appreciable ecological problem with such power stations.

Capital investment also will be much lower as compared to the conventional hydro-power stations. These are generally categorized under Mini-Micro hydro projects.

In India, a few bulb sets are operating : Teesta, (West Bengal, 4×7.5 MW), Sone-Link (Bihar, 5×1.65 MW).

5.2.6 Certain Other General Information About Hydro-power Stations

(A) In India, and in many other countries, hydro-power generation comprise of a substantial percentage of total installed capacity. Against a total estimated potential for around 1,50,000 MWs in this sector in India, so far (Dec., 2018), an installed capacity of 49521 MWs (including small hydro-stations: 4558 MWs) could be achieved, and that is around 15% of total installed capacity of

3,33,550 MWs (till Dec., 2018). Plans are in hand to add more capacity in this sector in the coming decade.

In some countries near 100% installed generating capacity comprise of hydro-power; Zaire-100%; Norway-99%, Brazil-83%.

(B) Advantages of Hydro-Power Generation

- Comparatively lower cost of generation at the project stabilized stage;
- Near zero pollution problem, there is no appreciable emissions;
- Mostly located away from populated areas;
- Require much less manpower to operate the station as compared to STG plants.
- Added advantages of flood control and irrigation facilities at some locations.
- Since this is a predominantly cold working system operating with a much lower rpm, maintenance problems are much lower as compared to Thermal power plants.

Fig. 5.1(A): A Typical Hydro-Power Station

(C) Some Drawbacks can also be Mentioned

- The project is dependent on vagaries of changing climatic conditions effecting availability of sufficient water in some years/seasons,

- High capital cost and long gestation periods,

- Certain ecological problems which can crop up at the project site/surrounding areas even after many years of operation (mostly hydrological problems).

    >>A generalized schematic layout of a Hydro power generating station is given in Figures No. 5-2. **

    ** NOTE: This layout is based on a hydro station with Francis Turbine having a valve in the penstock system.

(D) Depending on the penstock length and size, the head and the discharge rate, there can be a Surge Tank[+] installation connected to the penstock system; also there can be more than one valve (disc or spherical or both same type) placed in the penstock.

NOTE: Some times when the valve in the penstock is suddenly or quickly closed due operational reasons, the flowing water in the penstock experience a back-pressure surge, the energy of which can be tremendous and same has to be dissipated without causing damage to the penstock. The surge tank connected to the penstock at a suitable location can absorb this surge by giving a vent through the surge tank (normally the top of the surge tank kept open to atmosphere).

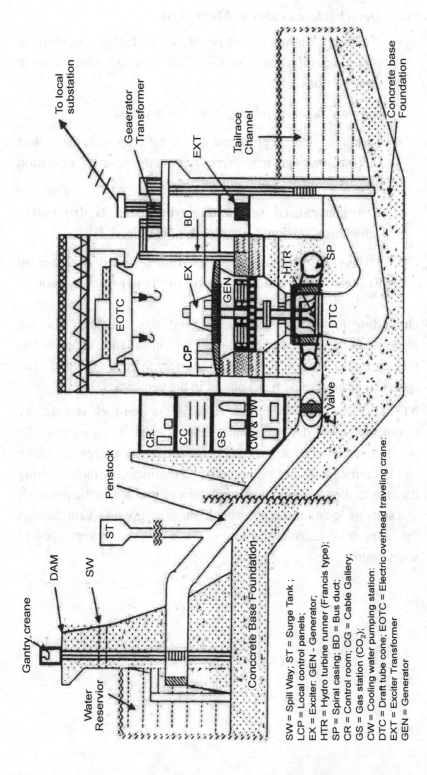

SW = Spill Way; ST = Surge Tank ;
LCP = Local control panels;
EX = Exciter: GEN - Generator;
HTR = Hydro turbine runner (Francis type);
SP = Spiral casing; BD = Bus duct;
CR = Control room; CG = Cable Gallery;
GS = Gas station (CO_2);
CW = Cooling water pumping station:
DTC = Draft tube cone; EOTC = Electric overhead traveling crane:
EXT = Exciter Transformer
GEN = Generator

Fig. 5.2: Schematic Diagram of A Generalised Layout of A Hydro-Electric Power Station (A Vertical Sectional View) (Not to Scale)

5.2.7 Ecological Issues Connected with Hydro-Stations

(A) hydro-power stations, specially large ones with a back up reservoir can cause:

   (*i*) Inundation of large areas up stream affecting the population living in the catchment areas (resulting in displacements) as well as forestry and cultivable agricultural lands going underwater (Example: old Tehri Town in Uttarakhand).

   (*ii*) During construction, blasting of rocky mountainous areas can trigger frequent land slides since due to repeated blasting, soil and rock formations become unstable.

   (*iii*) During rainy seasons due to heavy rains or cloud-bursts (which is common in high mountainous areas), can cause high flood situations down stream (controlling gates on the spill-way) may have to be opened up fully during high floods to avoid the reservoir water level rising beyond safe limit).

   (*iv*) Due to dams and diversion channels, some areas in the down stream zones can become draught prone affecting agricultural and fishing activities rather severely, particularly during dry seasons.

(B) All the above issues will have to be considered while preparing the project plans (DPR) and suitable remedial measures incorporated in the construction plans to the extent predictable/assessable at that point of time.

(C) It has been the experience that over the decades, that rivers start changing their flow behaviour and course thus affecting the performance of the hydro-power stations. Also, due to climate change, rain flow patterns as well as ice melting rates/glacial conditions in the upper reaches in the mountains changes thus appreciably causing changes in the water flow conditions; many dam project gets starved of water flows particularly during dry seasons and hence the performance of the existing hydro-generating sets goes down; one or two of the installed sets may have to stopped due to low water level in the reservoir.

(D) By and large, hydro-stations are environment friendly when compared with thermal or nuclear power stations.

6

Nuclear Power Generation Plants

> *"Nothing is lost in the Universe; Matter turns into Energy, Energy turns into Matter"*
>
> — *Gautama Buddha,(around 600 BC).*

6.1 NUCLEAR POWER GENERATING PLANTS

Nuclear Power Generating Plants are a class by themselves. In a Nuclear Power Station, heat is generated due to continuous liberation of energy from the atomic fission chain reactions taking place in the core of the Reactor Vessel where radio-active fissile fuel rods of natural Uranium (U-238), or enriched Uranium (U-235), depending on the type of reactor, are placed in multiple packets in an inner vessel made of special grade stainless steel/titanium alloys called the Calandria. The prefabricated fuel rods (U-235) is processed in a fuel enrichment plant(separately located), extracted from natural Uranium (U-238); these rods are individually inserted inside highly polished Zirconium (Zircaloy-2) or zirconium coated titanium alloy tubes and then this assembly is again put inside another such tube of bigger diameter leaving annular space in between for passing of coolant (water, heavy water, CO_2 etc.) (double casing is for isolating uranium rods from the coolant as the same corrodes the uranium rods).

6.2 THE REACTOR CORE

The Reactor core is placed inside a sealed steel reactor vessel (Calandria), fabricated from special grade thick steel plates, formed and welded by very stringent technological processes and quality control procedures. The core is shielded by a thick layer (about a meter thick) of graphite brick-lining along the inside walls of the reactor vessel. On the outer side of the reactor vessel body, a biological shielding is provided, first with a metre thick annular column of water and then the whole reactor body is encased in a concrete building with wall thickness of around three metres.

6.3 USE OF THE HEAT FROM THE REACTOR CORE

The heat from the reactor core is evacuated by means of a coolant or moderator cum coolant fluid circulated by specially designed pumps and pipelines in a hermatically enclosed loop system; the heated fluid is then passed through a special type of Heat Exchanger called HRSG (Heat Recovery Steam Generator). Inside the HRSG, besides the hot fluid, water flows through a secondary circuit of metallic tubes (stainless steel tubes/titanium tubes), completely isolated from the hot radioactive fluid-flow circuit. The heat from the hot fluid gets eventually transferred to the water flowing through the metallic tubes (as mentioned) and the water gets eventually converted into superheated steam, which in turn gets fed to the steam turbine that rotates a electrical generator to generate electrical power, just like the system prevailing in the conventional STG* thermal power plant. Besides this moderator-cum-coolant fluid (which can be Light Water, Heavy Water or gas like Helium, CO_2, N_2 or even sodium vapour depending on the design and type of reactor), a set of boron-carbide rods are also kept partially inserted in the reactor core for moderation and control of the chain reaction process within the safe/permissible limits. The reactor vessel is rested on heavy steel and concrete foundation lined with graphite bricks.

At the top, the end-shield cover of the reactor vessel is also made of special grade thick steel plates and lined inside with graphite bricks to prevent escape of radioactive gases/fine radio-active particulate matters.

(*STG = Steam Turbine Driven Generator).

The selection criteria for the coolant or the moderator-cum-coolant fluid plays a very crucial role in the reactor design and its operation; in fact, the reactors are classified based on the use of this fluid, such as: Light Water Reactor (LWR)/Boiling Water Reactor (BWR), Pressurized Water

Reactor (PWR), Pressurized Heavy Water Reactor (PHWR), Heavy Water Moderated Pressurized Water Cooled Reactor (HWPWR), Gas Cooled Reactor (GCR), Advanced Gas Cooled Reactor (AGR), high Temperature Gas Cooled Reactor (HTGR) etc.

NOTE: India has adopted the PHWR version with natural Uranium (U-238) as the fuel for most of its Nuclear Power Stations, whereas France has adopted mostly gas cooled-gas-moderated ones. (Gas Cooled Gas Moderated predominantly uses CO_2 for many advantages). CANDU (Canadian Deuterium Uranium) type Heavy Water Power Reactors are also being used in many Nuclear Power Plants. **Figure 6.1 presents a set of schematic diagrams of different types of nuclear reactors for power plants.**

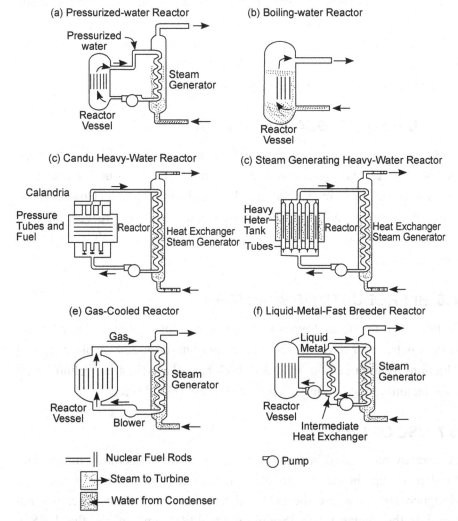

Fig. 6.1: Diagrams of Different Types of Nuclear Reactors for Power Plants

The above present schematic configurations of various types of Reactors some of which are adopted by the manufacturing companies in conformance with customer preferences.

6.4 FAST BREEDER REACTORS

There is another category of Nuclear Reactors being developed - called the Fast Breeder Reactor (FBR) which can create new fissile materials even from non-fissile materials as it operates and consumes the initially fed fuel of U-235 or U-238 or even Plutonium-239. These are still in advanced stage of development but yet to be commercially exploited (may become a reality in the coming few years). In fact India has already developed and successfully commissioned one 13 MWe FBR set at Kalpakkam. A 500 MWe set is under development. Once this system becomes commercially operative, It will revolutionize Nuclear Power Generation process (more dependable techno-economic safer option).

6.5 FURTHER RESEARCH IN INDIA

Further research is going on in India for using a mixture of U-233 (isotope of U-238) and Thorium in the next generation FBRs to utilize country's vast reserve of Thorium deposits available in Kerala-Konkan Coastal areas; this may become a reality in the coming decades but there are technological hurdles to overcome, hence some uncertainties persist (each types has its unique features).

6.6 BI-PRODUCTS OF REACTORS

In fact, even the normal types of reactors mentioned above, also create as bi-products, other fissile materials mostly Plutonium-239 (A bi-product of Uranium fission reaction process) which has other scientific/technological uses including the making of Atomic Bombs (If so desired).

6.7 USE OF HRSG

As already mentioned in paragraph 6.3 above the moderator-cum-coolant fluid picks up the heat from the reactor, gets pumped into a specially designed HRSG where the hot fluid transfers the heat to water flowing through the metal tubes (Stainless Steel/titanium) inside the HRSG.

This process eventually transforms the water in to super-heated steam at temperature around 220°C-300°C, at pressure 45 to 70 ATA; this steam is fed to the steam turbine-generator set installed in a adjacent building, slightly away from the Reactor building. The HRSG is also located in a separate building adjacent to the reactor building. The rest of the equipment and process used and followed for electrical power generation are more or less like the ones used for the conventional STG-thermal power generator set described in **Chapter 3**.

6.8 MODERATING/COOLING FLUID

Since the moderating/cooling fluid flowing around the reactor core (and then circulating through the HRSG) is highly contaminated by radio-active emissions, it becomes very hazardous to handle and process. To have additional safeguards against this, an intermediate set of heat exchanger may be used where the hot radio-active moderator/cooling fluid transfers heat to another fluid like sodium vapour (or other gaseous fluids) circulated through the HRSG for transferring heat to the water flowing through metallic tubes placed inside the HRSG where steam gets generated, and as stated above, then this steam gets fed to the steam turbine generator set to generate electrical power. Between the intermediate Heat Exchanger and the HRSG, very special measures are incorporated to keep the radioactive contamination within the permissible limits. In this duplicate Heat Exchanger arrangement, obviously heat losses in the system will be more and therefore, the system will have comparatively lesser heat efficiency and also more costly to take care of higher degrees of biological safety.

6.9 STEAM PARAMETERS OF HRSG

As a result of the inherent process of indirect heat transfer to water, the steam generated in HRSG normally will have poorer parameters: such as super-heat temperature of 220°C–300°C, pressure:45 ATA to 70 ATA, as compared to the steam from the conventional boiler used in a conventional sub-critical STG Plant with temperature of 535-540°C at pressure 160-170 ATA. Such lower steam parameters impose substantial changes in the steam turbine design and configuration. Since temperature and pressure are lower, volume of steam required increases substantially for the same power out put; this in turn, for the same power output, increases the size of the turbine cylinders, blades, number of stages of blades (both moving and fixed). Also since normally, no Reheat process is there, IP gets abolished, or gets merged with HP; the LP may get multiplied in number depending on the total power-output desired. Conventionally, the size of the LP or any cylinder cannot be increased beyond certain practical limits because of certain associated engineering/technological constraints including those relating to transportation/handling and optimization of operation.; the last stage diameter of LP is kept around 4-meters and below, which is the universal practice with most manufacturers.

Compared to the conventional steam turbines for conventional STG thermal stations, steam parameters being poorer, the steam turbine for Nuclear sets will have lower overall thermal efficiency for the same range of power output.

A generalized schematic diagram of a Nuclear Power Station set-up is given at Figure No. 6.2.

6.10 THE ELECTRICAL GENERATOR

The electrical generator, being driven by the nuclear steam turbine, along with its auxiliary equipment and systems, will be similar to those adopted for conventional thermal power stations (STG Plants). As an universal practice Nuclear Turbine runs at 1500 or 1800 RPM (called 1/2 speed) and thus the Generator will have a 4-Pole Rotor configuration.

6.11 THE TURBINE OPERATING PARAMETERS

The turbine operating parameters will have to be necessarily matched to the designed operating parameters of the reactor and the HRSG, to avoid operational problems.

6.12 THE NUCLEAR TURBINES

The Nuclear turbines are conventionally designed and designated as Half-Speed machines, since their RPM is 1500 or 1800, depending on the system frequency desired for the Electrical power generation; this requires the generator rotor to have a 4-pole configuration.

This is also forced by the condition of poorer steam parameters delivered by the HRSG (as mentioned above).

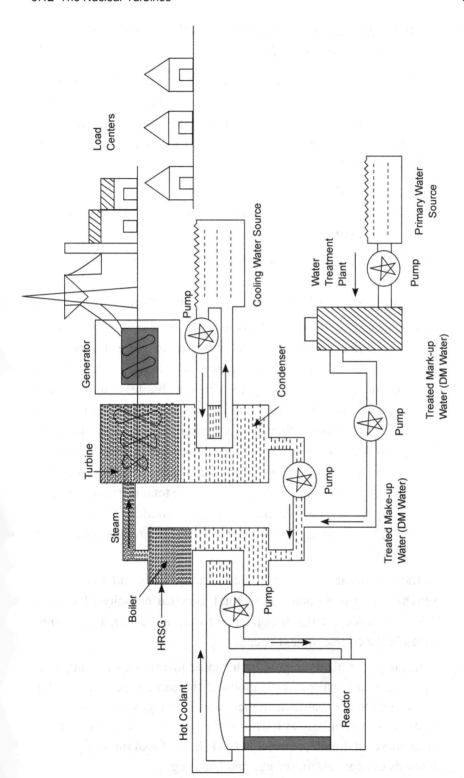

Fig. 6.2: Schematic Arrangement of A Fission-Reactor Nuclear Power Plant

6.13 OTHER IMPORTANT INFORMATION REGARDING NUCLEAR POWER PLANT

(*i*) Although nuclear power was attractive as a cost-effective alternative source of energy in 1950s/60s/70s (for augmenting the installed capacity by filling up, to some extent, gaps between demand and installed capacity available), over the last three decades, because of various factors like :

(*a*) Scarcity of fissile materials/uranium.

(*b*) Restrictions on export of nuclear materials and technology by most countries.

(*c*) Strong public opinion against Nuclear stations (because of perceived hazards and ecological problems, particularly after the Chernobil APS (Russia) mishap in 1986).

(*d*) Because of high cost of fissile materials and their processing/prefabrication, the nuclear power is no more as attractive and cost-effective as it were three decades ago. Nevertheless countries which can get sufficient fissile materials and operate their Nuclear plants at a cost-effective regimen, or countries where other fuels are much costlier or scarce, (and also because of ecological safety problems), they still prefer to go in for nuclear plants. In fact, there are countries where the percentage of nuclear power to the total installed capacity is quite substantial, to name a few: Sweden (51%), France (76%), Finland (56%) (Ref; UNO Energy Statistics Year Book).

(*ii*) In India, nuclear power installed capacity is around 6780 MWs (2018) which is around 2% of total installed capacity of around 3,33,550 MWs. (National objective: to achieve a target of 40,000 MWs in the course of next decade).

(*iii*) Nuclear power plants require high capital investments as compared to the conventional power stations of the same capacity. Nuclear power plants are viable at higher capacity ranges, say above 220 MW units, particularly at locations far away from sources of fossil fuels like coal, lignite, petroleum (NOTE: for fossil fuels, high costs towards transportation, storage and handling).

(*iv*) Nuclear power plants can be cleaner, as there is no big paraphernalia required for fuel storage, fuel supply and handling, ash and slag disposal etc. However, nuclear radio-active residues pose tricky problem for disposal/dumping for which suitable biologically isolated safe places are to be found, developed and maintained, to avoid ecological hazards.

(*v*) Only 12 kilograms of U235 can give a power output of 100 MW for a month as against thousands of tonnes of coal (30000 to 35000 T) required for a conventional STG thermal power plant to give that much power output for a month (depending on, of course, the quality and heat rate of the coal used).

(*vi*) The first nuclear power stations (5 MW) was commissioned for commercial operation in the then Soviet Union (now Russia) in 1954.

(*vii*) The largest single unit of nuclear power generating set of 1462 MWs is operating in KK-Isar-II APS in Germany (there are a number of sets operating in the world in different countries with unit ratings between 500 MW and 1462 MWs). In India, so far, the highest unit capacity nuclear set of 1000 MWs ratings (2 nos.) are operating in Kodan Kulam APS. 4-6 more such plants are in active planning for installation in the course of next 5-10 years.

(*viii*) Nuclear Power plants needs refueling (in reactor) a least once a year. Depending on design and operating cycle, refueling is done either by shutting down the unit for 2 months or so; but as per the latest technology adopted, refueling can be done on-line without shutting down the reactor, but this feature has to be incorporated in the reactor design stage incorporating adequate controlling systems for safe operation and disposal of spent fuel and other radio-active residues/ discharges.

(*ix*) Researches are going on in several countries (Russia, Japan, UK, France, USA, Germany) for the last more than four decades for the development of a controllable **Thermo-Nuclear Fusion Reactor** (the process which is going on all the time in the sun, liberating enormous amount of energy; it is estimated that earth's surface gets approx. 1/23, 000,000,0th part of that energy in the form of heat/light/cosmic rays/ultra-violet rays etc.; also the process in which Hydrogen-Bomb releases its enormous destructive energy).

The experimental set-ups for this are generally known as 'Tokamak', 'LHD (Large Helical Device)', 'MFER, (Magnetic Fusion Energy Research)' 'TFTR' etc. International collaborative efforts are also going on through International Thermonuclear Experimental Research (ITER) project involving USA, Russia, Japan and the European Union (EU) countries. India is also an active participant in this endeavour. Once this controlled Fusion Reactor comes to reality and commercialized (projected time frame for such an achievement is 2030-2050), the energy problem of the world can be solved for ever since for the fusion reactor, Deuterium (Heavy hydrogen: an Isotope of hydrogen) which is abundantly available dissolved in the sea water by natural process, is used as the basic fuel combined with Tritium (another isotope of hydrogen which is radio-active in nature and also found dissolved in sea water by natural process) (use of D-T fuel combination); it is estimated that the level of radio-active emissions in this process will be much less and more safer to operate.

6.14 LIFE-CYCLE OF NUCLEAR POWER PLANTS

Nuclear Power Plants (based on nuclear fission process) can have a life-cycle of over 30 years, depending on: (#)

  (*i*) Technological features of the Reactor (type, capacity rating, material of construction, configuration, fuel used).

 (*ii*) Operating cycle and loading pattern followed.

(*iii*) Periodical major overhauling done for critical parts/components, auxiliaries (including Replacements as required) at the prescribed operating periods as suggested by the manufacture (for reactor, steam turbine, generator and critical auxiliaries).

#REMARK: There are Nuclear sets that are running productively for over 35 years.

Renewable and New Energy Sources for Electrical Power Generation: Solar Power Plants Wind Turbine Driven Power Plants Co-Generation Power Plants Biomass Based Power Plants Geo-Thermal Power Plants Tidal Energy Based Power Plants Fuel-Cells for Electrical Power

"Human Beings are always on the look out for energy, both Internal and External. Right from the moment they arrive on earth, God has already made arrangements in various ways, to meet their energy needs to support their survival till the moment of their destined departure. Overuse and unethical exploitation of energy resources will surely make them hungrier all the time, leading to loss of physical and mental balances; greed always harms at the end."

— from a discourse given by a Spiritual Speaker during Kumbh Mela.

REMARKS:

(*i*) With increasing levels of ecological/environmental problems arising out of fossil fuel based power plants (which being the major contributors to the power generation capacity in most parts of the Globe), there is increasing emphasis on shifting to the Renewable Sources of Energy and electrical power generation from those sources. It is estimated that in the course of next two decades or so, a major quantum of Electrical Power will be derived from the Solar Energy, Wind Farms, and Geothermal Energy, besides increasing capacity in the field of Nuclear Power Generation although the same is fraught with certain degree of Ecological, Biological and Dual Use risks.

(*ii*)　Power generation prospects for large scale commercial generation from Fuel Cells are still years away since this technology is still in the experimental stage of development. With emerging technological innovations, this system may someday become a commercially viable option; presently, there are limited usages in certain defence equipment systems (sub-marines in particular) and on space vehicles.

7.1 SOLAR POWER PLANTS

7.1.1 Present Status

Solar Energy is an endless source, abundantly available in many parts of the Globe, specially during certain seasons depending on the geographical locations.

In India, solar energy can be harnessed almost through out the year. Based on the presently available economically viable technology, Electrical Power is derived though the deployment of arrays of the Solar Photovoltaic Panels placed at vantage locations where sufficient sunrays can be made use of almost the year round, 8 hours a day.

As per the present estimates (Source: Annual Report: 2015-16 of The Ministry of New and Renewable Energy, Government of India), the total potential of Solar Power Generation in the country, is around 7,48,990 MWs; present total installed Solar Power Capacity is around 17,052 MWs as in 2017: The National Objective is to reach a Solar Power capacity of 25,000 MWs in the course of the next 5 years.); coming decades may see a substantial growth in this field because of many advantages including the techno-economic viability, as compared to the other conventional non-renewable methods.

7.1.2 Solar Photo Voltaic Power Plants: Technological Concepts and Advantages

(i) Solar photovoltaic cells of certain technologically convenient sizes (101 mm round and 125 mm pseudo-square; thickness-325 microns) are fabricated from ultra-pure silicon based amorphous semiconductor materials; the fabrication process is carried out in a highly controlled dust- proof climatic conditions (with strict control on temperature and humidity) in special enclosures/ production labs. These cells have embedded electrical conducting circuitry (commonly made of Silver/Silver-Aluminum alloys, or other super conducting alloys).

Depending on the panel sizes to be made (as per the standards prescribed), these cells are arranged on the panels and are inter connected electrically; the inter connection of the embedded circuits will depend on the pre-decided voltage range and power output module to be achieved for the planned system. An array of such panels are placed at the selected site, facing the sun by proper orientation, and supported by appropriate structural arrangements. Matching power evacuation/supply control apparatus are installed at a suitable location at the project site (a local control room which in turn gets connected to the distribution/grid networks).

Figure No. 7.l(A) represents an Example of Solar Cells and **Figure No. 7.1(B)** represents an example of Solar Panel Arrays placed at a location.

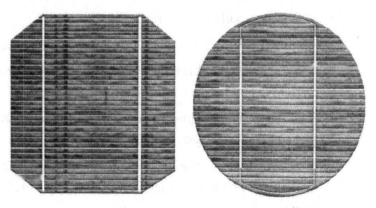

Fig. No. 7.1(A): Typical Solar Cells

100 kWp SPV Grid Interactive Power Plant at APTRANSCO, Hyderabad

Fig. 7.1(B): Typical Solar Panel Arrays

(ii) Solar Photovoltaic Technology enables direct conversion of sunrays to Electricity, in the Direct Current (DC) Mode, which eventually gets converted to Alternating Current (AC) Mode through appropriate Inverter Devices incorporated in the system. This process provides an attractive alternative option vis-a-vis the conventional sources of electricity for many advantages as listed below:

- Silent, non-polluting and renewable,

- Highly reliable with minimal maintenance efforts and costs,

- Can be in Modular arrangements and versatile for different system configurations at the same project site.

- Can be installed almost anywhere, with location-specific structural support systems..

- Cost-effective generation and distribution when compared with other conventional electricity generation and distribution systems.

7.1.3 Types of Solar Photovoltaic (SPV) Power Plants Presently in Vogue:

- Grid-Interactive System,
- Roof-Top System,
- SPV-Hybrid Systems with combinations as under:
  - PV-Mains,
  - PV-Diesel,
  - PV-Mains-Diesel,
- Standalone PV System,

In the following paragraphs, we are making brief presentations of the above systems for basic familiarization:

(A) Grid-Interactive Systems

(1) Grid-Interactive PV systems are connected to the utility grid network to enable wide area distribution of power to the proximity load centres. These state-of-the-art systems operate during the day time for 6-7 hours and deliver high quality power to the load-centres directly, being synchronized with the grid. During the periods of low load usage by connected proximity load centres, the excess power is exported to the grid for enabling other away centres to use the same. These systems can be provided with short-time Battery backup to feed power to the critical loads during grid outages.

(2) These systems can be used for power supply to:
- Industries,
- Farmhouses,
- Schools, Colleges, Hospitals,
- Irrigation pump feeders,

(3) Available in various capacities, from around 25KWp onwards, depending on the aggregate module size adopted.

Figure No. 7.l(C) represents a schematic layout of the Grid Interactive SPV System.

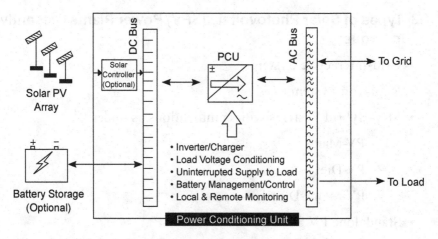

Fig. 7.1(C): Grid Interactive SPV System

(B) Roof-Top Systems

These systems are installed on roof tops of buildings for dedicated power supply to the building premises; capacity range can be 2 KWp onwards . These can be used to supplement the local load requirements; available in both single phase and three phase versions.

These systems are often visible in urban and semi-urban areas.

(C) SPV Hybrid Systems #

(1) The SPV-Hybrid systems are designed to operate in conjunction with both Grid system and Diesel generators, backup by Battery Bank system.

    The system can be used to supplement the load and can be configured either as a battery charger, or an inverter.

(2) The system can utilize a bi-directional inverter to regulate voltage supply to the load and functions as an on-line UPS. The inverter operates in parallel with either the grid supply, or the diesel generator with active power to the load. The built-in intelligent software ensures optimum loading of the DG set leading to better Specific Fuel Consumption (SFC). Alternatively, the system can be configured as Solar PV-Mains Hybrid systems also.

(3) Advantages of the system can be as under:

    • Provides clean and reliable power to load inspite of variations in Solar irradiation.

- Reduces the fuel consumption of the DG set by operating at optimum efficiency with load management.

- Reduces pollution levels.

- Enables trade-off of capital and operating costs by providing more optimal solutions.

- The system minimizes the capacity requirement of PV array and that of storage batteries.

- Provide fully automatic uninterrupted power output and full protection from power cuts.

NOTES:

(*i*) Batteries used in the system will have to be of very high quality with high level of reliability.

(*ii*) Functional health of the battery bank is critical and must get regular attention for good level of upkeep.

(*iii*) It is a Smart Automated system backed up by dedicated software systems.

Figure No.7.1(D) Illustrates the schematic layout of a typical SPV Hybrid system.

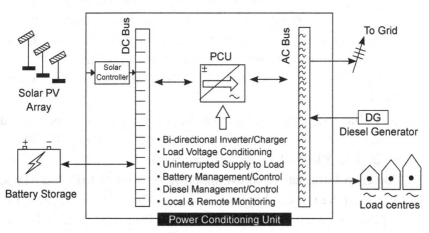

Fig. 7.1(D): A Typical SPV-Hybrid System

(D) Stand Alone SPV System

Stand Alone Solar PV power plants are those which are not connected to the utility grid. These power plants are primarily meant to cater to the needs of Rural Electrification in remote areas. System ranging from 1 KWp onwards with suitable battery banks and matching inverter device can be designed and installed at the desired locations.

Figure No. 7.1(E) represents a schematic layout of a typical Stand Alone SPV power station.

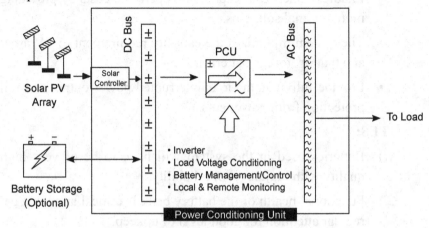

Fig. No. 7.1(E): Stand - Alone SPV System

7.2 WIND TURBINE DRIVEN POWER GENERATING PLANTS

(A) Wind Turbine driven power generation system has also emerged, in the last four decades or so, as a techno-economical environment friendly option for large-scale commercial power source. Starting with unit capacity of a few KWs in the initial stages, in some countries by technological innovations, the unit capacity has reached around 8000 KWs (Burbo Bank Wind Farm, UK.-2017).

In India, the larges unit capacity so far achieved is around 3,000 KWs. The installed capacity in India reached till 2018 is around 32,848 MWs (Grid-Interactive). The potential of Wind Power in India has been estimated to be around 1,02,772 MWs. Amongst the Renewable Energy Sources based power generation capacities, Wind power has taken the lead over others; by the year 2022, it is expected that wind power capacity will reach around 40,000 MWs.

(B) A Brief on Technical Parameters of Wind Turbine-Generators

    (*i*) Depending on the geographical locations and the perennial wind flow conditions, both overland and in coastal zones, the height of the tower may range from 50 metres to even

195 metres (Burbo Bank Wind Farm, UK,).The composite weight of the turbine-generator set, along with the propellers, can be from around 600 kgs to over 8000 kgs; added to this, occasionally, the lateral wind force during storms/cyclones, as well as the seismic forces during earthquakes, can be considerable.

All these factors will have to be considered while designing the towers and propellers for safe operation and a reasonable life-cycle; foundation and anchoring of towers will be important factors.

(*ii*) The length of the propeller blades can be from around 5 metres to even 50 metres, depending on turbine rotating capacity (torque generation and speed, supported by the high wind flow conditions as prevailing in UK's Liverpool Coast).

(*iii*) Both the turbine and the coupled generator are enclosed in a hermatically sealed high strength but light weight alloy steel/composite material capsule type enclosure. The speed governor and the exciter for the generator are also embedded inside the capsule. The set is remote-controlled from the ground/shore-based control station. Generated power is evacuated through insulated power cables (under ground/sub-marine depending on location) which are connected to the switch-gears/control apparatuses located in the ground station from where power is fed to the load centres/grid system.

(*iv*) Depending on the design adopted, the generator terminal voltage can range from 1.1 KV to 6.6 KV. System, with adopted system frequency can be 50 Hz or 60 Hz, depending on the system preferred.

(*v*) Mounting the set on the tower is a rather tricky affair, needing the services of Tower Cranes, Barge-mounted cranes, or even large helicopters, or both.

(*vi*) Periodical maintenance services as well as trouble shooting are challenging tasks; compatible methodologies have to be planned and incorporated while designing and executing the project.

(*vii*) Anchoring/braking/holding the propeller in stationary position during maintenance/troubleshooting, or during system outages, is again a tricky job; appropriately designed braking/holding device has to be incorporated in the system for this purpose.

(*viii*) Because of the possibility of widely variable speeds due to varying wind speeds, sensitive voltage regulation system has to be incorporated in the turbine-generator control apparatus.

(*ix*) To reduce the weight of the propellers, the blades are made of composite materials/carbon fibres which can withstand high degree of mechanical stresses (twisting, bending forces).

(*x*) A Hybrid system of combination with Battery Banks and Diesel GEN sets can be configured for stable/assured power supply in the vicinity load centres.

>> **A picture of a Typical Wind Farm is Presented at Fig. 7.2.**

7.3 CO-GENERATION POWER PLANTS

(*i*) These are mostly in the category of captive power generation plants which utilizes the surplus/bye-product heat emanating from other process industries such as sugar mills, petro-chemical and fertilizer units, pharmaceutical factories and the like. The heat transfer medium is generally steam with rather low parameters of temperature and pressure, emanating from a HRSG (temperature from 200–300°C; pressure: 40-50 AT A) placed in the system. To enhance the heating capacity, the HRSG can be fired one(based on furnace oil or gas).

(*ii*) The Turbine in this case is a single cylinder back-pressure type one (non-condensing) and drives a AC Synchronous 3-phase Electrical Generator; the turbo-generator set will have dedicated control apparatus and switch-gears with appropriate protection systems.

(*iii*) The unit capacity of such power plant may range 500 KWs to 5000 KWs depending on the quantum of steam (with requisite parameters) which the parent operating system can make available.

The generating set can be synchronized with the Main Bus System of the parent industrial unit for parallel operation and load sharing,

Fig. 7.2: A Typical wind farm

(*iv*) Since such cogeneration plant is primarily meant to be a captive unit, the capacity sizing of the unit may be limited according to the requirement of the parent industry. In rare cases, the surplus power may be sold to the proximity Grid network by contractual arrangements with the concerned utility company.

7.4 BIO-MASS BASED POWER PLANTS (#)

(A) These plants basically are of limited capacity thermal power units utilizing Steam generated in conventional boilers by burning different types of bio-mass that get generated as:

(*i*) Agricultural rejects such as buggas (dried sugarcane rejects from Sugar mills/Jaggery Units), dried rice and wheat hays/ stubs,

(*ii*) Municipal garbage (urban, semi-urban).

(*iii*) Rejects from food processing industries.

(*iv*) Rejected Molasses/effluents from the breweries.

\# **Bio-gas plants are excluded from this topic since bio-gas is not normally being used for commercial generation of Electrical Power because of certain limitations.**

(B) Steam generated in such boilers is of low parameters (temperature ranging from 200-300°C, pressure: 30-40 ATA), and therefore are not suitable for large-scale power generation: the unit capacity may not exceed 15MWs at one location.

(C) In case of such power plants, obnoxious emissions (Co, CO_2, NO_X, SO_X, H_2S) will be a big issue, and adequate emission control/ neutralization devices will have to be installed to meet statutory regulatory pollution control guidelines..

(D) The steam turbine in such plants will be a single cylinder machine; the coupled generator, a 3-phase AC Synchronous machine, for running at 1500/1800 rpm speed, or, 3000/3600 rpm as per the National Standards of system frequency followed.

(E) Such plants are run as supplementary power source during peak-load periods, but are not suitable for quick start and quick stop; therefore, their operation will have to be governed by rather tight operational scheduling matching with the rate of availability of fuels of the kind and the electrical loading pattern.

(F) Ensuring regular flow and the quality (specially with respect to calorific values obtainable) of the combustible fuel materials will always remain a challenge for the operating administration.

7.5 GEO-THERMAL POWER PLANTS

(A) Geothermal Energy, abundantly available in the Earth's core below the earth's crust/tectonic plates and originating from the super heated Magma present at the Mantle of the earth's core which is a perennial source of tremendous quantum of heat.

(**NOTE:** Scientists have estimated that earth's total heat content is about 10^{31} Joules, a part of which can be harnessed, by development and application of appropriate technological innovations, for generation of millions of MWs of Electrical Power in the long run).

With the presently developed technological processes, a part of this heat energy could be brought out on the earth's surface, at certain prospective locations, in the form of hot fluids (hot water, steam, hot gases), which in turn are used to generate steam of the required parameters suitable to drive a steam turbine to rotate a coupled electrical generator of the matching capacity.

The generated power is then transmitted to the proximity load centres; the same also can be connected to the proximity Grid systems.

(B) Deep bores, ranging from 1.5 km to even over 6 km, are drilled into the earth crust at certain prospective locations, to reach the geothermal heat source which is then brought up on the surface by means of insulated pipe lines in which the hot fluid flows upwards by means of natural or induced pressure; this hot fluid, which can be in the form of steam, or hot water, or gas, is eventually made to convert to usable quality steam to drive a steam turbine-generator set to deliver electrical power.

REMARK:

There are geographical indications on the earth surface at certain locations about the presence of hot water springs and natural Geysers which can be further explored/made use of by installing appropriate turbo-machineries to generate Electrical Power; the generation process is by and large similar to the one followed for a conventional STG plant.

(C) **Technological Processes Presently Adopted For Geothermal Power Generation**

Originating the idea of such power generation process in 1904 in Italy, the process has since then undergone certain technological innovations, eventually adopting three generally preferred processes as described under:

(1) **Dry Steam Power Stations**

These stations directly use the steam spouting from the hot spots/geysers at around 150°C or higher with sufficient pressure and discharge rate to turn steam turbine which in turn rotates a coupled electrical generator to deliver power to the connected load centres/grid. The exhaust steam is condensed in a cooling chamber from where the condensate water is pumped back to the source well below the earth surface; the cycle gets repeated as long as the power station is in operation. In case for some reason, the turbo-generator is put out of operation, the incoming steam may have to be discharged to the atmosphere since it may not be possible to hold the same in any accumulator(techno-economic issues involved).

Figure No. 7.5(A) represents a schematic arrangement for a typical Dry Steam Power Station.

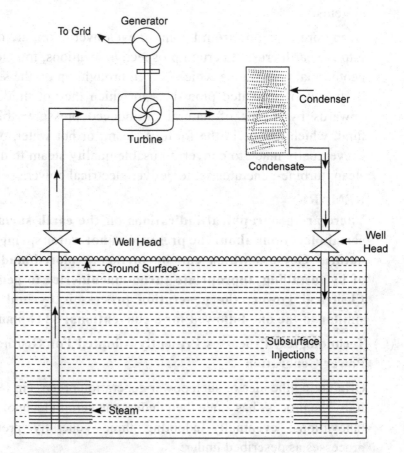

Fig. No.7.5(A): Dry Steam Power Station

(2) **Flash Steam Power Stations**

In this system, hot water from deep below the earth surface is pulled up by deep well pumps and then forced into lower pressure tanks where the hot water gets converted into flash steam which is used to drive steam turbines coupled with electrical generators (there can be multiple units depending on the quantum of steam that can be generated from the setup).

These turbines require the fluid/steam with temperature of at least 180°C or more, with requisite built up pressure.

In most cases, the hot water flows up at its own pressure through the Wells dug in the ground. As the hot water flows upwards, the pressure decreases and some of the hot water boils into steam in the low pressure conditions.

This steam is then separated from the water in a separator tank from where the steam is fed to the turbine to drive the electrical generator.

The left over water and condensed steam may be injected back into the underground source reservoir, making the set up a potentially sustainable power source.

Presently, Flash Steam Stations are the most preferred Geothermal Plants.

Figure No. 7.5(B) represents a schematic arrangement of a Flash Steam Power Station.

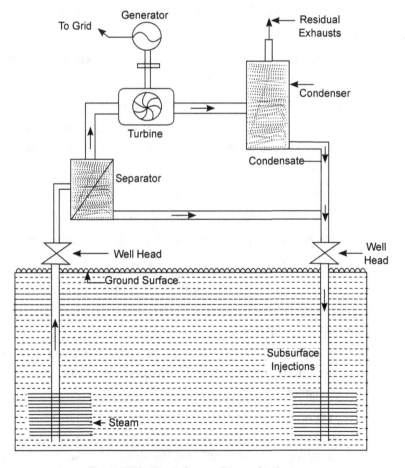

Fig. 7.5(B): Flash Steam Power Station

(3) **Binary Cycle Power Stations**

These are the latest technology Geothermal Power Stations developed recently.

This system can accept fluids at temperature as low as 57°C.

The moderately hot geothermal water is forced up mixed with a secondary fluid with a much lower boiling point than water (alcohol, methanol, methane gas, sodium vapour, as per the technological preferences). This causes the secondary fluid to flash vaporize which then drive the gas turbine-generator sets to deliver electrical power; the choice of the secondary fluid will be a critical factor keeping in view the problems related to operational sustenance, and associated techno-economic and environmental issues, on a long term perspective.

(D) **Advantages and Drawbacks of Geothermal Power Systems**

(1) **Advantages:**

- Abundant source of heat energy for a long term sustenance for power generation;
- Comparatively much lower level of harmful emissions;
- Much lesser number and size of infrastructural requirements vis-a-vis conventional STG/GTG Plants; hence lower capital costs.
- There is no Fuel requirements for the stations;
- The stations can operate in a clean environment;
- Such stations can operate with much lower manpower.
- The overall cost of generation per unit will be lower.
- Geothermal stations require minimal land area and fresh water.

(2) **Drawbacks Expected:**

- Prospective locations will be dependent on geothermal potentials of the place; prospecting costs can be quite substantial.
- There is uncertainties for long term sustenance; some wells can get depleted/dried up over a period, rendering the project unsustainable. It is rather difficult to estimate the life-cycle of the set up.

- In the prospecting stage, such issues will require very elaborate scientific investigations which is a costly affair and this will have a bearing on the project costs;

- Fluids dawn from deep earth carry a mixture of eco-unfriendly Gases such as CO_2, H_2S, Methane, Ammonia, Radon; these contribute to global-warming, acid rains, radiation hazards, and noxious smells if released without proper pollution control measures incorporated in the project infra-structure; this adds to the capital as well as to the operational costs;

- In addition to the dissolved gases, hot water drawn from geothermal sources may hold in solution traces of toxic chemicals such as mercury, arsenic, boron, antimony and salt; these chemicals come out of the solution as the water cools and can cause environmental damages if released untreated.

  The modern practice is to inject such geothermal fluids back deep into the earth to stimulate production; as a side benefit this helps in controlling the adverse affects on environment.

- The construction of project on certain locations can adversely affect Land stability, cause subsidence of land mass which can trigger earthquakes.

- Geothermal power stations can also disrupt the natural cycle of Geysers in some locations.

- For some existing geothermal stations, it has been estimated that the station can produce an average 45 kg of CO_2 equivalent emissions per MW-hr of generated electricity, as compared to coal fired stations emitting around 1001 kg of CO_2 equivalent per MW-hr.in case appropriate emission control devices are not installed.

(E) **Present Status of Geothermal Power Generation**

As many as 24 countries have installed geothermal stations (2016 status) with about 12636 MWs total generation capacity (highest capacity of 3450 MWs is in USA); more units are expected to come up in the next decade to reduce dependence on Fossil Fuel based stations which have substantial adverse impacts on environment.

In India so far no such geothermal station exists; plans are in hand to develop its first geothermal station in Chhattisgarh state; the project is in the exploratory stage.

International Geothermal Association (IGA) provides techno-economic assistance to its member states.

REMARK:

Interested readers may refer to Wikipedia, the Free Encyclopedia, for more details on the subject; Also refer to the Annual Report of the Ministry of New and Renewable Energy, Govt, of India for information on the Indian Scenario.

7.6 TIDAL ENERGY BASED POWER GENERATION SYSTEMS

(A) Oceans, covering almost 70% of earth's surface, represent an enormous amount of energy demonstrated in the form of waves, tidal surges, marine currents and thermal gradients that get created due to inter-continental movements of the vast water masses. There is an enormous potential to harness this energy to generate electrical power through the use of appropriate technology and conversion devices.

Scientists have estimated that there are potentials in the World for generating 49000 MWs from Waves and Tides, and another 1,80,000 MWs from the Thermal Gradients of the oceans

(B) Although the idea was mooted many decades ago, the development of an appropriate and commercially exploitable technology is still in the nascent stage, may require a good deal of experimentation and time before finally adopting a dependable/consistent/viable technology for sustainable commercial power generation. A variety different technologies are under development in different parts of the globe. As an eco-friendly alternative source for large scale power generation, this will remain an attractive option in the coming decades.

7.6.1 Tidal Power Stations

(A) This option is presently in the priority consideration for such hydro-energy based power generation and many countries including India, are actively working on such projects for further development in order to eventually give the same a practical shape in the coming years. Although not yet widely explored and used, Tidal Energy has the potential for future large scale generation of Electrical power.

(B) The tidal cycle occurs every 12 hours due to the gravitational forces of the Moon (and other celestial bodies including the Sun) exerting gravitational pulls on the water mass of the seas causing a bulge in the water level causing in turn, a temporary increase in the sea level; when the sea level rises, the water from the middle is forced to rush towards the shorelines creating the tidal surges.

The difference in sea/ocean water heights from the Low Tide to the High Tide is the potential energy which can be harnessed to generate electrical power similar to the traditional process of hydro power generation from dams. The tidal water can be captured in a barrage across an estuary during the high tide and forced through a hydro-turbine during the low tide. In order to generate sufficient quantum of power from the tidal energy potential, it has been estimated from experimentations that the height of the high tide must be at least five metres higher than the low tide,

(C) There are only around 20 coastal locations on the earth with tides of such heights, and one of them is in India; the Gulf of Cambay and the Gulf of Kutch in Gujarat on the west coast have the maximum tidal height of the range of 11 metres and 8 metres, with average tidal range of 6.77 m and 5.23 m respectively, an workable range.

Both the State and the Central Government have joint plans to develop such power stations in this zone with location-specific technology. The Gulf of Kutch in Gujarat can have a 50 MWs station in the immediate phase, extensible to 200 MWs at a later date.

(D) According to the estimates of the Government of India, the total potential for tidal power in India is around 8300 MWs, which includes:

- 7000 MWs from the Gulf of Cambay,
- 1200 MWs from the Gulf of Kutch,
- 100 MWs in the Gangetic Delta of Sunderban region of West Bengal.

(E) **Tidal Energy can be an Attractive Options** in the area of the Renewable Energy Sources since, unlike the Solar and Wind energy, the occurrence of tidal cycles can be predicted quite accurately through out the seasons for enabling operational planning for dependable power outputs.

(F) There are a number of technological options under consideration a few of which have taken practical shape with varying degree of operational success. More advance models of submersible hydro-turbine-generator sets are under development for working both, during the up coming high tides as well as during ebb/low tide conditions: that is the turbine can work during flow of tidal waters in both directions (tides coming in, tides going out).

Figure No. 7.6 depicts the schematic arrangements of a Typical Tidal Power Station using the concept of a Tidal Barrage in the Tidal Basin created in the prospective coastal location.

(G) **Advantages and Disadvantages of Exploitation of Tidal Energy**

As in the case of most natural phenomena based projects, the Tidal Energy based ones also have their share of Advantages and Disadvantages, as listed below:

Advantages

- Tidal Energy is renewable energy source, is free and clean, requires no fuel and no waste bi-products are generated.
- Tidal energy has the potential to produce a great deal of free and green power.
- Tidal power stations are much cheaper to install, operate and Maintain.
- Low noise pollution as any sound generated largely absorbed by the surrounding water mass.
- High degree of predictability as high and low tides can be predicted years in advance, unlike wind and sunrays.
- Tidal barrages provide protection against flooding and land damages.
- Large tidal reservoir have multiple uses and can create recreational lakes in the areas where there was nothing existed earlier, thus creating additional income sources and indirect employment opportunities.

Disadvantages:

- Tidal energy is not always a consistent energy source as the same is dependant on the strength and flow of the tides which themselves are effected by the gravitational effects of the moon and the sun (which can have limitations under certain circumstances not controllable by human beings).

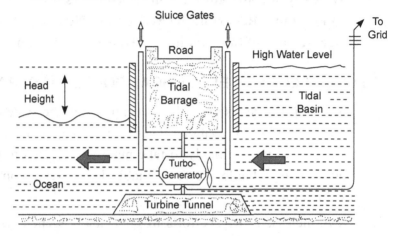

Fig. No. 7.6: Tidal Barrage Power Plant

- Exploitation of Tidal Energy requires a suitable site where the tides and tidal streams are consistently strong enough.
- The infrastructures created for harnessing the energy for power generation must be such that they can with stand the forces of nature, and this will result in high capital and maintenance costs.
- High costs towards power evacuation/distribution, for laying (and maintaining) the underwater cables from the submerged generating sets.
- The system can generate power only for a limited period (around ten hours a day) during high and ebb tides.
- Can cause changes in estuary eco-system and an increase in coastal erosions where the tides are concentrated for the purpose of the projects.
- Build up of silts, sediments and pollutants within the tidal barrage areas from the proximity rivers and streams flowing into the basin as these are unable to flow out into the sea freely.

- Danger to fish and other marine lives getting stuck in the barrage, or sucked into the submerged turbine blades; appropriate preventing measures will have to be installed around the project sites which lead to additional capital costs.

REMARKS: Interested readers may refer to Wikipedia, the Free Encyclopedia, for more details on the subject; Also refer to the Annual Report of the Ministry of New and Renewable Energy, Govt, of India for information on the Indian Scenario.

7.7 FUEL CELLS FOR ELECTRICAL POWER GENERATION

(A) Although the idea is more than a century old, the system is still under evolution for enabling lager scale commercial generation of electrical power; likely to take many more years to develop appropriate technologies for a viable larger scale commercial exploitation.

Nevertheless, the R&D efforts are on by many agencies including BHEL in India.

(B) **Historical Background**

(*i*)   One scientist named William Grove in 1839 first discovered that it is possible to generate electricity by adopting a process of reversing the electrolysis of water that normally takes place in the traditional voltaic cells the precursor of modern day batteries.

(*ii*)   In 1889, two researchers, Charles Langer and Ludwig Mond, advanced the idea of '**Fuel Cell**' by using air and coal gas put in a container subjecting the mixture to chemical reactions to produce electricity through ion-exchange principles.

(*iii*)   In early 1900s, attempts were made to develop Fuel Cells that could convert coal or carbon into electricity by induced chemical reactions in a container with suitable electrolytes.; however this proposition did not succeed at that point of time for practical application since in the contemporary period, **Internal Combustion Engine** came into existence with a roaring success and became more attractive as a prime-mover for delivering power, both mechanical and electrical. The idea of fuel cell, however, was not abandoned, and the research efforts continued in one form or the other.

Figure No. 7.7(A) presents the basic principle of a Typical Fuel Cell;

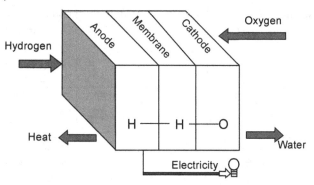

Fig. 7.7 (A): Principle of Operation of a Typical Fuel Cell

Figure No. 7.7(B) shows the Major Sub-systems of Typical Fuel Cell Power Plant.

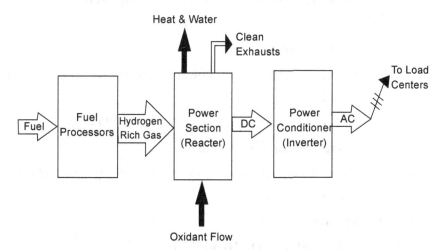

Fig. 7.7 (B): Major Sub-Systems of a Typical Fuel Cell Power Plant

(C) Present Status of Fuel Cell Developments

There are a number of technological options under active pursuits, with varying degree of success. The major technological models that are in advanced stage of development are the five mentioned in the Table below:

Table 7.7 (X)

Fuel Cell Type	Operating Temp. (°C)	Unit Size (kw$_e$)	Most likely Applications
1. Alkaline Full Cell (AFC)	70–100	<100	Space and Military
2. Proton Exchange Membrane Fuel Cell (PEMFC)	50–100	1–100	Portable devices like Lap Tops, Cell phones, Video cameras, Domestic Appliances, Buses, Cars RLY, Locos, Tram cars.
3. Phosphoric Acid Fuel Cells (PAFC)	160–210	5-200 Also, MW Size plants.	Dedicated power(+heat), Railways.
4. Molten Carbonate Fuel Cell (MCFC)	650	100–2000	Dispersed power and Utility Power (up to 100 MWs)
5. Solid Oxide Fuel Cell (SOFC)	800–1000	upto 100 MWs	Domestic and Commercial Power and heat., CCPP, Mobile power for Rlys.

(D) **Some Leading Engineering/Technology Agencies Active in the Field**

D-l. SOFC PLANTS

(*i*)   Siemens Westinghouse Power Corporation, USA, for SOFC Plants.

(*ii*) Fuel Cell Technologies, Canada (SOFC Plants).

(*iii*) NKK of Japan (SOFC Plants).

(*iv*) Rolls Royce Allison & MC Power (SOFC and Hybrid GT Plants Combination).

(*v*)   In India:

   • BHEL (R&D), Hyderabad.

   • Central Glass and Ceramic Research Institute, Kolkata.

   • IISC, Bangalore.

D-2. MCFC PLANTS

   • M/s. Fuel Cell Energy Inc, USA/Canada.

   • M/s ANL fuel Cell Group, Denmark.

- GE Energy & Environmental Research Corp., USA.
- Hitachi, Japan.
- MTU, Germany.
- NERF, Netherland.
- In India, M/S. Central Electro-Chemical Research Institute, Karaikudi.

D-3. PAFC PLANTS

- M/S International Fuel Cells Corpn.,USA, in a Joint Venture with M/S United Technology (USA) and M/S Toshiba (Japan).
- Desert Research Institute, Nevada, USA.
- Fuji Corporation., and Mitsubishi Electric Corporation, Japan.
- In India.

  > BHEL-R&D, Hyderabad.

  > Naval Mineral Research Lab., Ambernath.

D-4. PEMFC PLANTS

- M/S. Ballard Power Systems Inc., Canada, USA.
- In Japan, M/S. Ebara Ballard Corprn., Mitsubishi, Toshiba, Toyota Motors.
- M/S Anuvu Inc., Avista Labs., BCS Technologies and others in USA.
- DeNora, Itally.
- Xcellsis, Germany.

D-5. AFC PLANTS

- M/S Astiris Energy Inc., Canada.
- Apollo Energy Systems, USA.
- In India, CECRI, Karaikudi.

CONCLUDING REMARKS:

(*i*) **Advantages:** Silent and Eco-friendly power generation system, requiring much less space and auxiliaries, easier to maintain and service.

(*ii*)   The above listing show that there is lot of interest in the field being pursued by many Technological Organizations in various countries,

(*iii*)   In the coming decades, it is expected that many break-throughs will be made for viable commercial power generation from the Fuel Cell based systems.

❑❑❑

8

Miscellaneous Related Topics That Seek to Provide Supplementary Knowledge for Power Station Management

> *"When you endeavour to acquire a larger knowledge structure, then you must also take care of the smaller elements that go into supporting that bigger structure to be sturdy and durable".*
>
> — *Author's view.*

8.1 ENERGY TERMS AND DEFINITIONS

Please read ANNEXURE-III for details and try to understand the significance of each of these terms.

8.2 PARALLEL OPERATIONS OF ELECTRICAL POWER GENERATING PLANTS

In most modern day power stations, there are more than one Turbine-Generator units installed and run at the same time in the same power station#; in fact in the same turbine hall for convenience of operation and control, besides saving in space and infrastructure.

(# **NOTE:** *Exception may be in Nuclear Power Stations for certain valid reasons*).

In order to add to the output capacity and enabling load-sharing, the units are put in parallel operation mode. To facilitate satisfactory and stabilized operation, there are certain preconditions that are to be met before the units are put into the parallel operation mode; these can be enumerated as under:

(1) The Generators must have the same No-Load Terminal Voltage ratings;

(2) The Percentage Reactance should be same (or nearly same) to avoid flow of mutually circulating currents (which increases internal loading of the stator windings of the generator having the higher %-age reactance which causes depletion in the terminal voltage of that generator; it also causes undesirable internal power loss, lowering the efficiency of the machine.

(3) The designed/actual output frequency rating of the generators should be same (variations of upto +/−1.5% can be allowed with adequate controlling mechanisms in place to avoid power angle instability, specially during wide load variations (both in magnitude and quality) on the common station bus system.

(3) The terminal phasing should be same, i.e, R-R, Y-Y, B-B; N-N

(N-N are inter connected through the Neutral Grounding system); this is ensured while connecting the generator terminals to the common Station Bus System through it's dedicated transformer and circuit breaker;

(4) Matching generator protection system must be incorporated in the generator control panels (protection relays: over-load, reverse-power/anti-synchronism, differential protection, earth faults, over-voltage, terminal/bus short-circuiting, bus duct faults, Neutral break/disconnect, lightning protection).

(5) The Capacity Ratings of the units should be nearly same, should not be too widely different in value. In case of widely different capacity ratings, there may be difficulty in maintaining synchronism/system stability when load sharing is attempted during parallel operation; there can also be reflections on the turbine performance of the respective units; the Turbo-Supervisory apparatus may find it difficult to respond to such widely different loading patterns.

(7) Vector Groups and the Power-Factor ratings of the generator should be matching to allow unified operations, to avoid difficulties in synchronized loading within the stability regimen.

(8) Generator Stator Insulation Class should be same or compatible to avoid di-electric imbalances and losses arising out of higher current density in the over-hang bends of the stator windings and resultant temperature rise beyond permissible limits. For large generators, generally Insulation Class-F, which allows a temperature rise of upto 155°C in the stator windings, is adopted.

NOTE: The recognized classes of insulating materials and the Temperature range assigned to them are as follows:

Class	Temperature(°C)
Y	90
A	105
E	120
B	130
F	155
H	180
C	Above 180

There are a variety of insulating materials developed to meet specific characteristics;

Electrical Engineers are supposed to get adequately familiarized with such insulating materials, including their di-electric properties.

(9) The Excitation System of the generators should be of the same type with compatible control systems.

(10) The direction of rotation of the turbo-generators should same for enabling unified operational control.

(11) Generator Transformers and dedicated Switch-gears and circuit-breakers Must be compatible for sustained regimes of parallel operation. The **Short-Circuit Ratings (SCR)** of the Circuit-breakers are determined by experts based on the magnitude of the Fault Current flows expected during various fault conditions that can occur on the connected station bus systems or on the connected proximity grid network. (most severe fault condition occurs when all three phases are short-circuited and earth-faulted simultaneously; the magnitude of the SCR is to be based on such fault conditions).

(12) Installation of Proper Earthing and Lightning Protection systems, for the generators, generator transformers, circuit breakers and the entire station switch-yard be ensured.

NOTES:

I. *Neutral point of large generators (mostly with STAR connected in their stator windings; can also have Double-Star Winding layout) is firmly earthed though an Earthing Transformer; this facilitates monitoring of the magnitude of any earth-leakage current and detecting the extent of system unbalances that may occur under different loading conditions on different phases.*

II. *It is to be kept in mind that the generators are connected to the station bus system for parallel operation through the dedicated Generator Transformers; therefore, these transformers also have to work in parallel subject to their fulfilling certain conditions which are akin to the ones mentioned above at paragraph 8.2 (1-8,11,12).*

8.3 POWER FACTOR (PF) MANAGEMENT

This is an important aspect of Economic/Cost-effective Management of Power Plants with respect to input-output energy balancing, and is broadly a measure of healthy operating conditions.

Although the recommended PF range is 0.85-0.90 lagging for satisfactory performance of the system, practically it becomes rather a difficult proposition because of the varied loading conditions:

(*i*) Most industrial loads are Inductive in nature (A large number of induction motors are deployed and this brings down the PF to even below 0.6 adversely affecting the system; under Such Condition, active component of the power consumptions as recorded in the **kwhr meter** is much lower than the actually consumed **kvahrs**, thus causing financial loss to the power station. The actual cost of fuel/energy consumption by the turbine-generator set for delivering the demanded **kvahrs** will not be fully realized from the consumer because under the low PF condition in the system, the actual kwhrs registered in the meters will be much lower.

To remedy this anomaly, power company insists on the consumer to install power factor correction equipment, mostly Capacitor Banks of appropriate ratings .

NOTE: *The PF issues beyond the scope of the power station limits, that is in the connected grid network, is NOT being covered in the current discussion; for a large grid system, capacitor banks alone will not be enough; there are other technological solutions like incorporation of super-synchronous AC generator running at Leading PF is installed and connected at a vantage point in the grid network system; this generator is not normally put on load and just float on the network, but it consumes some power which is traded off against the benefits of improved PF in the system.*

(*ii*) Some foundries use induction furnaces and electric arc furnaces which tend to lower the system PF substantially; such consumers are also advised to install Capacitor Banks of appropriate ratings,

(*iii*) Another source of low PF is the use of a large number of Tube Lights and fans in large offices, hotels, hospitals, universities and the like; they are also advised to install appropriate Capacitor Banks to improve the PF to the prescribed value.

(*iv*) The Generators and the Transformers installed in the power station can also inject inductive current in the system tending to lower the system PF. This aspect is taken care of by installing Capacitor Banks of appropriate ratings in the station switch-yards.

(*v*) The above corrective steps will require regular checking and vigilance on the part of the concerned engineers of the Power Company.

8.4 VOLTAGE REGULATION OF THE GENERATORS

This is an important aspect of quality of generated power under variable loading conditions: with high magnitude of load current demand on the generator, the generator tends to slow down, (effect of reverse torque on the generator shaft as well as on the coupled turbine shaft) having a drooping effect on the terminal voltage which cannot be allowed to have a variation of more than +/–5% (or lower) of the rated value. Under such circumstances, the installed Voltage Regulator comes into operation by way of:

(1) The Generator demanding and getting increased excitation current from the Excitation System to maintain the level of magneto-motive force (flux density) required to compensate for any depletion in

the terminal voltage, thus maintaining the terminal voltage at the desired/preset value.

(2) At the same time, sending a signal to the turbine controls to:

(*a*) **In Case of STG Plant:***

Admit additional steam into the turbine to meet extra torque demand on it's rotor shaft because of higher load coming on the generator; the process is automated for continuous operation. To respond to variations of steam demands by the turbine, the boiler also has it's matching controls to adjust ingress of DM Water, firing level in the fire chamber and generation of steam as well as the steam delivery system. Sometimes due to rather quick/drastic fall in the steam demand by the turbine, the extra steam already present in the boiler system is blown off automatically to the atmosphere, and the boiler eventually stabilizes to the required steam supply condition.

(*b*) **In Case of Hydro-power Plant:***

In the similar operational situation, the station control system sends a signal to the turbine controls to increase the water inflow into the hydro-turbine runner to deliver enhanced torque to the hydro-turbine rotor shaft to maintain the prescribed rotational speed required for stabilizing the hydro-generator terminal voltage to the desired/specified value; the process is automated for continuous operation.

(3) **Voltage Regulators/Stabilizers** are designed to respond to any appreciable variations in load demand on the generator which tends to cause variations on the speed, in turn on terminal voltage.

(4) **A Reverse of (1) and (2-a, b) above** can take place if the load demand on the generator comes down appreciably. The sensitivity level of the control systems is a critical factor and the system designers will have to be adequately conversant about this aspect while designing /adopting the voltage regulation system; a number of options are available depending on the capacity rating of the unit; but universally, the regulation is effected by controlling the magnitude of the excitation current in to the rotor windings of the generator.

* **REMARK:**

    Similar process for Voltage Regulation in case of other types of power plants are also adopted, but the technology and the configuration may be different to meet unit /project specific requirements.

8.5 POWER TARIFF SYSTEMS (#)

(A) The business of power generation and distribution is normally supposed to be a techno-commercial endeavour, and therefore, investments (capital and working capital funding) made on power projects must bring in a reasonable Return on Investments (ROI) to keep the enterprise viable on a long term perspective (the life-cycle of a power unit/station can be over 30-35 years with proper efforts on upkeep and maintenance).

(B) Electricity is not a commodity that can be stored in large quantum and used as and when required. A flat rate based on units (kwhrs) consumed was not found to be satisfactory in view of quality of loading, magnitude of loads and the ruling PF conditions with respect to large consumers.

    It was then decided to adopt a Two-Tariff system of charges to be realized from the consumers, specially those whose consumption quantum is above certain prescribed threshold level.

(C) **The Two-Part Tariff system** is based on two costs of which the first part is covered by an annual or monthly/quarterly amount, and the second part is based on the price of units consumed.

    The **Load Factor** of a consumer over a prescribed period also has a bearing on the two-part tariff system cost assessment.

(D) The two-part Tariff usually are of two different types: one for Industrial Consumers, and the other for Domestic Consumers.

    An Industrial Two-part Tariff is always based on the Maximum Demand,-either in kW, or in kVa, recorded in the meters installed. Rates to be charged for kW hrs consumed will be the second part of tariff and meters separated.

(E) The Electricity Tariffs are ultimately based on the National Energy Tariff Policy guide-lines issued by the Authorized Agency and the same are binding on the Power Companies.

However, certain minor variation may be allowed in case of certain power projects installed on PPP model to attract Capital Investments in the field (PPP = Public-Private Partnership).

(#) REMARK:

The Power/Energy Tariff is a substantive subject and the concerned engineers will have to undergo specialized training or advanced study courses on the topic; there can be a number of variable algorithms involved.

8.6 MANAGEMENT/ADMINISTRATIVE ISSUES IN PESPECT OF POWER PLANTS

All Power Plant Engineers must have a reasonable level of knowledge and proficiency about/on the following issues:

 (*i*) **Project Management Aspects:** Preparation of project reports, project planning and scheduling, project financing, budgeting and budgetary control methods.

 (*ii*) **Materials Managements Aspects:** Identification of requirements, drawing up of specifications for procurement, raising of indents, classification and execution of tendering process, storage and distribution, and also understand the significance of the applicable processes of Inventory Management and Control System.

 (*iii*) **Human Aspects** of project management and HR Affairs.

 (*iv*) Safety and Security Aspects.

 (*v*) Applicable National and International Standards.

 (*vi*) A reasonable, level of familiarization with process of Operations Budgeting and related Budgetary Controls; Get familiarized with the Significance of Managerial Economics.

 (*vii*) Applicable Statutory Regulations including Pollution Control/ Environmental Issues.

 (*viii*) A reasonable level of knowledge of the provisions of Electricity Acts and Rules, Boiler Act, Explosives Act, Factories Act, Applicable Labour Laws.

 (*ix*) Be reasonably clear about your assigned role and responsibilities. In case of confusion, try to sort out with the boss at the earliest; do not sit over unsolved issues; delays may complicate the matter.

 (*x*) Take advantage of all Training Opportunities coming by your way;

"IN ORDER TO GROW AT THE WORK PLACE, YOU SHOULD BE A FIRST TRACK LEARNER".

Young engineers must inculcate the habits of reading relevant books/literature to keep themselves up-to -date with the emerging developments in the field; while the Internet mostly gives information, it does not provide the level of knowledge inputs required for a successful corporate career.

(*xi*) Get a reasonable level of knowledge on the principles of the functioning of Electrical Power Transmission & Distribution Systems since the Power Stations are supposed to deliver services through these Networks.

8.7 SOME OTHER USEFUL ISSUES FOR POWER PLANT ENGINEERS

Some other issues that Power Plant Engineers may find useful for satisfactory professional practices:

(A) Engineers must have a reasonable level of knowledge of Engineering Materials and their characteristics. Electrical and Mechanical Engineers should also have a reasonable Knowledge of Metallurgical characteristics of metals they select to use in the design and construction of Power Generating Machinery and related structures.

(B) Both Electrical and Mechanical Engineers engaged in the design and construction of power equipment must have a reasonable level of appreciation of the role of each other. Electrical Engineers must keep in mind that manufacturing shop activities and the on-site activities are predominantly mechanical process oriented. Three is an imperative need for mutual role appreciation and cooperation for unified operational success.

(C) Also get reasonably familiarised with the Civil Engineering aspects of a Power Plant structural construction and related maintenance issues.

(D) Be reasonably educated and conversant with applicable Electronics systems and Computer Software, take help from the concerned expansion and when required.

Annexure-I

POPULATION, AREA, INSTALLED ELECTRIC POWER CAPACITY, AND PER CAPITA CONSUMPTION OF POWER IN VARIOUS COUNTRIES

(*Rank as per population (from Wikipedia-UNO); other data from UNO,IEA-PD-Files);

RANK*	COUNTRY	AREA IN SQ MILES	POPULATION (2016)#	INSTALLED POWER CAPACITY (MWs)	PER CAPITA POWER CONSUMPTION (Kwhrs)
	World		7,32,2811468	–	–
1	China	3706386	1373541000	1646000	2674
2	India	1269338	1266883598	333550	1000+
3	United States	3718691	323995528	1052863	12077
4	Indonesia	741096	258316051	55000	754
5	Brazil	3286420	205823665	135000	2516
6	Pakistan	310401	201995540	24820	405
7	Nigeria	356667	186053386	12520	128
8	Bangladesh	44598	156186882	8600	294
9	Russia	6592735	142355415	248000	7481
10	Japan	145882	126702733	313000	7371
11	Mexico	761602	123166749	65450	1932
12	Philippines	115830	102624209	19000	643
13	Ethiopia	435184	10374044	2400	65
14	Vietnam	127243	95261021	39000	1312
15	Germany	137846	80722762	204000	7371
16	Egypt	381660	94666993	38000	1510
17	Iran	636293	82801633	77000	2632
18	Turkey	301382	80274604	70000	2574
19	Congo, Democratic Republic	905563	4852412	2600	185
20	Thailand	198456	68200824	40000	2404
21	France	211208	66836154	129000	6448
22	United Kingdom	94525	64430428	96000	4795

Data beyond 2016 not available in the website; + 2017 status. These data go on being revised from year to year; consult websites of CEA, Power Ministry and IEA (Geneva) for latest.

RANK*	COUNTRY	AREA IN SQ MILES	POPULATION (2016)#	INSTALLED POWER CAPACITY (MWs)	PER CAPITA POWER CONSUM PTION (Kwhrs)
23	Italy	116305	62007540	120000	4692
24	Myanmar	261969	56890418	4300	193
25	South Africa	471008	54300704	46000	3904
26	Korea, South	36023	50924172	100000	9720
27	T anzania (including Zanzibar)	364898	52482726	1200	95
28	Colombia	439733	47220856	16000	1270
29	Spain	194896	48563476	102300	4818
30	Ukraine	233089	44209733	56000	3234
31	Kenya	224961	46790758	2281	162
32	Argentina	1068296	43886748	36000	2643
33	Algeria	919590	40263711	16000	1216
34	Poland	120728	38523261	40340	3686
35	South Sudan	728215	12530717	3700	55
36	Uganda	91135	37578876	711	110
37	Canada	3865081	35362905	137000	14930
38	Iraq	168753	38146025	28000	1101
39	Morocco	172413	33655786	7700	861
40	Afghanistan	250000	33332025	600	141
41	Venezuela	352143	30912302	31000	2523
42	Peru	496223	30741062	12000	1268
43	Malaysia	127376	30949962	30000	4232
44	Uzbekistan	172741	29473614	13000	1628
45	Saudi Arabia	756981	28160273	66000	9658
46	Nepal	54263	29033914	856	194
47	Ghana	92456	26908262	2800	341
48	Mozambique	309494	25930150	2600	462
49	Korea, North	46540	11134588	10000	1347
50	Yemen	203849	27392779	1533	189
51	Australia	2967893	22992654	67000	9742
52	Taiwan	13892	23464787	48700	10632
53	Madagascar	226656	24430325	500	53
54	Cameroon	183667	24360803	1100	250
55	Syria	71498	17185170	8200	989
56	Romania	91669	21599736	24000	2222
57	Angola	481351	20172332	1700	401

RANK*	COUNTRY	AREA IN SQ MILES	POPULATION (2016)#	INSTALLED POWER CAPACITY (MWs)	PER CAPITA POWER CONSUM PTION (Kwhrs)
58	Sri Lanka	25332	22235000	3400	494
59	Cote d'Ivoire	124502	23740424	1522	244
60	Niger	489189	18638600	102	64
61	Chile	292258	17650114	23000	3739
62	Burkina Faso	105869	19512533	300	61
63	Netherlands	16033	17016967	32000	6346
64	Kazakhstan	1049150	18360353	19000	4956
65	Malawi	45745	18570321	353	102
66	Ecuador	109483	16080778	6300	1305
67	Guatemala	42042	15189958	4139	586
68	Mali	478764	17467108	600	80
69	Cambodia	69900	15957223	1400	256
70	Zambia	290584	15510711	2300	709
71	Zimbabwe	150803	14546961	2200	549
72	Senegal	75749	14320055	1000	209
73	Chad	495752	11852462	41	16
74	Rwanda	10169	12988423	100	38
75	Guinea	94925	12093349	500	74
76	South Sudan	400367	12530717	255	55
77	Cuba	42803	25115311	6280	597
78	Greece	50942	10773253	19000	4919
79	Belgium	11787	11409077	21520	7099
80	Tunisia	63170	11179995	4600	1341
81	Czech Republic	30450	10644842	22000	5636
82	Bolivia	424162	10969649	2200	683
83	Portugal	35672	10833816	19000	4245
84	Somalia	246199	10817354	81	27
85	Dominica	18815	73757	3800	1223
86	Benin	43483	10741458	171	
87	Haiti	10714	10485800	313	38
88	Burundi	10745	11099298	66	36
89	Hungary	35919	9874784	9289	2182
90	Sweden	173731	9880604	39000	12853
91	Serbia	34124	7143921	7594	3766
91A	Kosovo	—	1883018	1600	1533
92	Azerbaijan**	33436	9872765	7400	2025
93	Belarus	80154	9570376	9200	3448

RANK*	COUNTRY	AREA IN SQ MILES	POPULATION (2016)#	INSTALLED POWER CAPACITY (MWs)	PER CAPITA POWER CONSUMPTION (Kwhrs)
94	United Arab Emirates	32000	5927482	28000	16195
95	Austria	32382	8711770	24220	8006
96	T aj ikistan	55251	8330946	5500	1440
97	Honduras	43278	8893259	2100	595
98	Switzerland	15942	8179294	19000	7091
99	Israel	8019	8174527	16250	7319
100	Papua New Guinea	178703	6791317	900	441
101	Jordan	35637	8185384	4200	1954
102	Bulgaria	42882	7144653	12130	4338
-	Hong Kong	NA	7167403	13000	5859
103	Togo	21925	7756937	86	141
104	Paraguay	157046	6862812	8820	1413
105	Laos	91428	7019073	3400	555
106	El Salvador	8124	6156670	1792	925
107	Eritrea	46842	5869869	100	51
108	Libya	679358	6541948	8900	1421
109	Sierra Leone	27699	6018888	100	33
110	Nicaragua	49998	5966798	1200	739
111	Denmark	16639	5593785	14010	5720
112	Kyrgyzstan	76641	5727553	3900	1920
113	Slovakia	18859	5445802	8095	5207
114	Finland	130558	5498211	16000	14732
115	Singapore	267	5781728	13000	8160
116	Turkmenistan	188455	5291317	4275	2456
117	Norway	125181	5265158	33800	24006
118	Costa Rica	19730	4872543	2900	1888
119	Lebanon	4015	6237738	2320	2565
120	Ireland	27135	4952473	9100	5047
121	Central African Republican	240534	5507257	44	36
122	New Zealand	103737	4474549	9722	8939
123	Congo, Republic of	132046	8133105	500	114
124	Georgia	26911	4928052	3718	1988

** including nagorno-Karabakh

RANK*	COUNTRY	AREA IN SQ MILES	POPULATION (2016)#	INSTALLED POWER CAPACITY (MWs)	PER CAPITA POWER CONSUM PTION (Kwhrs)
125	Palestine	NA	4356290	240	550
126	Liberia	43000	4299944	27	69
127	Croatia	21831	4313707	4400	3933
128	Mauritania	397953	3677293	519	217
129	Panama	30193	3705246	2700	2105
130	Bosnia and Herzegovina	19741	3861912	4310	2848
131-	Puerto Rico	NA	3578056	6100	5310
131	Oman	82031	3355262	8200	7450
132	Moldova	13067	3510485	434	1226
133	Uruguay	68039	3351016	4400	2984
134	Kuwait	6880	2832776	16000	19062
135	Albania	11100	3038594	1895	2564
136	Lithuania	25174	2854235	3900	3468
137	Armenia	11506	3051250	4100	1671
138	Mongolia	603905	3031330	1000	1847
139	Jamaica	4244	2970340	1000	942
140	Namibia	318694	2303315	500	335
141	Qatar	4416	2258283	8800	15055
142	Macedonia	9781	2100025	2057	3314
143	Lesotho	11720	1953070	80	409
144	Slovenia	7827	1978029	3370	6572
145	Latvia	24938	1965686	3000	3459
146	Botswana	231803	2209208	895	1674
147	Gambia	4363	2009648	91	149
148	Guinea-Bissau	13946	1759159	39	17
149	Gabon	103346	1671711	415	488
150	Trinidad and Tobago	1980	1220479	2353	7456
151	Bahrain	257	1378904	3900	18130
152	Estonia	17462	1258545	3138	65151
153	Swaziland	15942	1451428	200	1033
154	Mauritius	788	1348242	1100	1928

RANK*	COUNTRY	AREA IN SQ MILES	POPULATION (2016)#	INSTALLED POWER CAPACITY (MWs)	PER CAPITA POWER CONSUMPTION (Kwhrs)
155	Cyprus	3571	1205575	1700	3234
156	Timor-Leste	5641	1261072	NA	99
157	Fiji	7054	915303	300	874
-					
158	Djibouti	8880	846687	100	472
159	Guyana	83000	735909	400	1087
160	Equatorial Guinea	10830	759451	500	120
161	Bhutan	18147	750125	1614	2779
162	Comoros	838	794678	22	51
163	Montenegro	5415	644578	900	4343
-	Western Sahara	102703	587020	58	142
-	Macau	NA	597425	500	7532
164	Solomon Islands	10985	635027	37	124
165	Suriname	63039	585824	412	3243
166	Luxembourg	998	582291	2000	10647
167	Cape Verde	1557	553432	100	542
-					
168	Malta	122	415196	620	4817
169	Brunei	2228	436620	777	8625
-					
170	Bahamas	5382	327316	600	4888
171	Maldives	116	392960	77	763
172	Belize	8867	353858	200	1130
173	Iceland	39768	335878	2800	50613
174	Barbados	166	291495	200	3087
-	French Polynesia	NA	285321	200	2453
-	New Caledonia	NA	275355	600	7263
175	Vanuatu	4710	277554	30	201
176	Sao Tome and Principe	386	197541	20	329
177	Samoa	1137	198926	45	502
178	Saint Lucia	238	164464	88	1824

RANK*	COUNTRY	AREA IN SQ MILES	POPULATION (2016)#	INSTALLED POWER CAPACITY (MWs)	PER CAPITA POWER CONSUMPTION (Kwhrs)
-	Guam	NA	162742	600	9217
-					
-					
179	Saint Vincent and the Grenadines	150	102350	47	977
-	Virgin Island,	NA	34232	44	2921
180	Grenada	133	111219	51	1798
181	Tonga	289	106513	17	436
182	Micronesia, Federated States of	271	104719	18	1705
183	Aruba	NA	113648	320	7039
184	Kiribati	313	106925	7	260
185	Seychelles	176	93186	100	3219
-					
186	Andorra	NA	85660	520	6565
187	Dominica	181	73757	33	1223
-	Bermuda	NA	70537	168	8506
-	Cayman Islands	NA	57268	100	10477
-	Greenland	NA	57728	96	5196
-	American Samoa	NA	54194	41	1845
188	Saint Kitts and Nevis	101	52329	64	3821
-					
189	Marshall Islands	4577	73376	.52	8177
-	Faroe Islands	NA	50456	NA	5945
-					
190	Monaco	1	45233	NA	NA
191	Liechtenstein	62	37937	250	35848
"	Turks and Caicos Islands	NA	51430	100	3888
192	Saint Marino	24	33098	NA	NA
-	Gibraltar	NA	29328	43	6819

RANK*	COUNTRY	AREA IN SQ MILES	POPULATION (2016)#	INSTALLED POWER CAPACITY (MWs)	PER CAPITA POWER CONSUM PTION (Kwhrs)
	Virgin Islands, British	NA	34232	44	2921
193	Palau	177	28341	44	NA
-	Cook Island	NA	9556	9	3308
-					
-					
-					
194	Nauru	10	9591	5	2424
195	Tuvalu	NA	10051	6	NA
-	Saint Pierre and Miquelon	NA	5595	NA	7479
-	Montserrat	NA	5267	NA	4061
	Saint Flelena, Ascension and Trisanda Cunha	NA	7795	7	1193
-	Falkland Islands	NA	2931	10	4759
-	Niue	NA	1190	1	3126
196	Vatican City	17	1502	NA	6500

Annexure-II

ENERGY SOURCES OF INDIA

1. **Electrical Power Generation Installed Capacity (#)**

    (As on 31-12-2017)

    (*i*) Thermal :

Fuel		MWs
Coal	=	1,92,972
Gas	=	25,150
Diesel	=	838
Total	=	2,18,960

@Renewables include small hydro units, solar, wind, bio-mass, waste Heat recovery units;

(*ii*)	Hydro	=	44963 MWs
(*iii*)	Nuclear	=	6780 MWs
@ (*iv*)	Renewables	=	62847 MWs
	G.Total	**=**	**333550 MWs***

(*v*) Total Electricity Generation (energy) (2016-17) = 1200 (approx.) Billion kwhrs.

*NOTE: it is estimated that the countrywide total Demand at present (2017) is around 3,50,000 MWs, leaving a gap of around 20,000 MWs between demand and availability. The demand is estimated to go up to 3,60,000 MWs by 2022, i.e. addition of around 30000 MWs over and above the present installed capacity. At the present average estimated cost of ₹ 5.5 crores per MW, a capital investment of around ₹ 1,65,000 crores is required during the next 5 years (average ₹ 33000 crores per year).

#Source: Website of the Ministry of Power, Govt. of India. Consult websites of the Power Ministry and CEA for updated status.

2. Fossil Fuel Resources: (as in 2016) ($)

2.1. Coal and Lignite

(*i*) Estimated Proven Reserves in the country

- Coal   =   293 Billion Tonnes;
- Lignite = 42 Billion Tonnes;

  Total   =   335 Billion Tonnes*

  *   Presently recoverable = 69 Billion Tonnes;

(*ii*) Present production/ extraction (2016-17)
                                = 570 Million Tonnes,

(*iii*) Present consumption level for energy generation (2016)
                                = 500 Million Tonnes;

(*iv*) It is estimated that at the present rate of consumption, India's coal reserves (proven) may last for about 300 years.

(*v*) Indian coal is generally of poorer quality with around 24% Ash and Shale contents, and the rate of energy yield (thermal) ranges from 1600 Kilo-calorie/kg to around 2400 Kilo-calorie/Kg, as compared to high-grade coal from Australia having an energy yield of around 4000 Kilo calorie/Kg.

(*vi*) Around 20 million tones of high grade coal are being imported every year From Australia and some other countries, being partly utilized in Thermal Power Plants and the rest for Steel Plants (mainly coking coal for steel plants).

2.2 Oil (Petroleum): ($)

(MMT = Million Metric Tonnes). Bbl = Billion Barrels;

(*i*) Estimated reserves (crude) (2016)   = 5.675 Bbl;

(*ii*) Present consumption level (2016)

   (*a*) Crude Oil (In Terms of Refinery Throughput)
                                                = 225 MMT

   (*b*) Petroleum Products   = 160 MMT

(*iii*) Production (2016).

   (*a*) Crude oil   = 40 MMT

   (*b*) Petroleum Products   = 221 MMT

(*iv*) Imports and Exports (2016):

    (*a*) Gross Imports (crude and other Petro-Products)

                                         = 190 MMT

    (*b*) Exports (petro-products other than crude)   = 66 MMT

    (*c*) Net Imports; (2016)

        * Crude   = 125 MMT

        * Other processed petro-products   = (–) 18MMT

                                      Net = 105 MMT

2.3. Natural Gas : ($)

    (*i*) Estimated reserves in the country (2016)

                                   = 50400 Billion Cu. mtr;

    (*ii*) Gross Production (2016)   =   37 Million Cu. mtr;

    (*iii*) Total Annual utilization(2016)   =   49.5 Million Cu. mtr;

NOTES: $

    (*i*) To meet increasing demands for Electrical Energy, a substantial quantity of natural gas is being planned for import from Iran and Some other countries in the next decade.

    (*ii*) The above Data have been taken from the Websites of CEA, Ministry of petroleum, Government of India.

3. Hydro-power Potential

As per latest estimates (2017) potential for hydro-power generation is around 1,50,000 MWs, of which so far (30-11.2017) only around 49521 MWs could be harnessed (32%).

4. Nuclear

The present status is indicated at para 1 above ; the constraints are both technological and lack of availability of usable fissile materials (mainly uranium at present). With NSG- clearance recently granted and with the prospect of Indo-US Civilian Nuclear Agreement and similar agreement signed with Australia recently coming to operation, the prospect of large expansion in Nuclear power generation capacity in the next decade or so may become a reality (plans are in hand to add 40,000 MWs in next 10 years).

5. **Renewables**

The present status is indicated in para-1 above; plans and efforts are in hand to harness these alternative sources as much as possible in the next 10-20 years, specially Solar and Wind energy..

6. Total Electricity Generation (2016-17) =1200 Billion Kwhrs.

❏❏❏

Annexure-III

ENERGY TERMS AND DEFINITIONS

Coal: is the most abundant fossil primary energy source and the classic fuel for power and heat generation. The importance of coal will be further increased in the future by gasification and liquefaction processes.

Coal Equivalent (CE): is the reference standard for the energy content of various fuels. 1 CE or 1 kg CE represents the mean energy content of 1 kg of hard coal having a calorific value of 7,000. 1 TCE = 29.3 GJ.

Combined-Cycle Power Plant: The combination of gas turbines and steam turbines in fossil-fired power plants provides advantages in respect of short construction periods, financing, flexibility, economic efficiency and environmental compatibility both where new projects or retrofitting is concerned.

Combined Heat and Power (Cogeneration): The simultaneous supply of electricity and process heat or district heat by a power plant. Utilization factors of more than 80% can be achieved.

Efficiency: Efficiency is a measure of effectiveness of an energy conversion process, i.e. the ratio of the useful energy output to the total energy input.

Power: Power inherent or induced capacity of a system (devices, gadgets, plants etc.) to do measurable work (or deliver force to do work over a period of time to unleash energy).

Electric Energy: Electric Energy is defined as the electric power used over a specific period of time. If a 100 watt electric bulb burns for ten hours, the electric energy consumed is 1 kwh. (power × time = energy consumed).

Emission: Emission are defined as releases of a pollutant into the environment, e.g. gases, dust, waste heat, noise, vibration.

Energy: Energy is defined as the capacity of a system to produce an external effect (work, heat, light). As a matter of principle of physics, energy can only be converted from one form of energy into another. For electrical energy, the unit of measure is the kilowatt-hour (symbol; KWh).

Energy Programme: Goals set by local authorities, an industrial enterprise or a government to assure the supply of energy as well as measures to be taken to achieve these goals.

Energy Recovery: Utilization of energy remaining (waste or residual) as a by-product, e.g. of production processes, ventilation and air conditioning or in households. Frequently, it is technically feasible and attractive from the economic point of view.

Forms of Energy: in the scientific sense, a distinction is made, for example, between mechanical, electrical, chemical and magnetic energy, heat and nuclear energy and in the practical sense between primary energy, derived energy, final energy and useful energy.

Gas Turbine: A gas turbine is a device which involves fresh-air compression, hot-gas production and the generation of rotational movement in one casing, (basically an IC- Engine).

Generator: The mechanical energy of a drive shaft is converted to electrical energy in a generator using the dynamo-electric principle.

Gross Calorific Value: The gross calorific value of fuels indicates the energy liberated during complete combustion of a unit of mass of a fuel and the recovery of water vapour condensation heat.

Instrumentation and Control: In power plants measurements, control and monitoring are known collectively as instrumentation and control.

Petroleum: is a naturally occurring mixture of low or high viscosity liquid consisting mainly of hydrocarbon derivates. Through selected distillation and cracking processes in the petroleum refinery, it is separated into fractions by billing range, density and viscosity; it is a fossil fuel extracted from liquid gas reservoirs.

Quality of Power and Energy: In the energy and power plant sector is expressed in terms of time, function and cost (period of availability of stable supply, reliability of supply at prescribed parameters, cost of supply etc.)

Rated Output: The rated output of a power plant is the maximum continuous rating for which the generators have been designed.

Solar Energy: Predominantly originates from nuclear fusion reaction going on within the sun. Commercial solar energy utilization is made difficult by a low energy density and a strongly fluctuating insolation time. Presently, solar energy is converted directly to electrical energy by using photo-voltaic plates exposed to sun rays.

Sources of Energy: are defined as materials or systems that contain useful energy , e.g hard coal, uranium, water, fossil, gas/ fuels, solar rays, wind flow, falling water, burning wood.

Steam Turbine: This is a thermal machine converts the thermo-dynamic energy contained in steam to mechanical energy contained in steam to mechanical energy in the form of rotor revolution.

Load Ranges: The time dependent power demands of the ultimate loads on the system require a power plant operating mode adapting in terms of time. Depending on their operational and economic characteristics, power plants are operated for 24 hours, for a number of hours at a time, or merely for a short time during the day, i.e. base-load, intermittent load and peak-load operations.

Maximum Capacity: The maximum capacity of a power plant is defined as the maximum continuous capacity, limited by the lowest capacity part or system, and is determined by measurements during operation.

Natural Gas: is the fossil primary energy source extracted from the underground hydrocarbon fields with the highest increases in demand. Besides the economic aspects, environmental protection favours the utilization of natural gas in power generation.

Nuclear Power Plants: are thermal power stations utilizing the nuclear fuels uranium, plutonium and thorium with releasing combustion gases. Heat generated in the reactor due to atomic fission process is utilized in a HRSG to generate steam which drives seam turbines which in turn drives an electric power generator.

Oil Equivalent (OE): is the referenced standard for the energy content of various fuels. 1 kg OE represent the mean energy content of 1 kg of oil having a calorific value of 10,000 kilo-cal. 1 TOE = 41.9 GJ.

Hydro-Turbine: This is a machine which converts the kinetic energy of falling water into mechanical rotating energy which in turn drives a generator for generating electric power.

Usage Factor. The usage factor of power plants is defined as the average utilization of the maximum capacity within a specific time, measured in hours per annum (h/a).

Plant Load Factor (PLF): This is defined as the level of capacity utilization of a power plant with respect to its designed/installed capacity

Utilization Factor: The utilization factor of a power plant is defined as the average operating efficiency, taking into consideration the part load behaviour.

Source: Siemens AG Pocket Diary 1998

❏❏❏

Annexure-IV

QUANTITIES AND UNITS

The term quantities designates physical and technical materials characteristics of objects and processes. Base quantities physically independent of one another.

Base Units of the International System of Units (SI)

(As defined in "Law of Units in Measurement")

Quantity	Unit	Symbol
Length	Meter	M
Mass	Kilogram	Kg
Time	Second	S
Electric current	Ampere	A
Thermodynamic Temperature	Kelvin	K
Amount of Substance	Mole	Mol
Luminous Intensity	Candela	Cd

Definition of Base units 9 Considerably Simplified in Part):

- Meter is the distance covered by light in a vacuum in 1/299792458 second.

- Kilogram equals the mass of the international prototype Kilogram.

- Second equals 9,192,631770 periods of the radiation of an atom of Cesium-133.

- Ampere is the intensity of a current which, by flowing through two indefinitely long, parallel current- carrying conductors, exerts a force of 2/10 Newton/m upon this pair of conductors.

- Kelvin is equal to 1/273. 16 of the absolute temperature of the triple point of water

- Mole is the amount of substance of a system which contains as many elementary entities as there are atoms in 0.0112 kg of Carbon-12.
- Candela is the luminous intensity in a given direction of a radiant source which emits radiation of a frequency of 540×1012 hertz and has a radiant intensity in that direction of 1/683 watt per unit solid angle.

Selected Derived Units and Terms

- Differences in temperature are expressed in Kelvin
- Temperature in degrees Celsius/°C equals thermodynamic temperature/K-273.15
- Temperature range 1°C equals 1 temperature range 1 K.

Temperature in degrees Fahrenheit/°F equals 9/5 (temperature in degree Celsius/°C) + 32; 1.8° F equals 1°C.

Name of Unit	Unit symbol	Definition	Previous unit Symbols
Becquerel	Bq	1 Bq = l/s	1 Ci = 37 G bq
I Hertz	Hz	1 Hz = 1/s	
Joule	J	1 J = 1 N x m	
Newton	N	1 N =1 kg x m/s^2	
Pascal	Pa	1 Pa = 1 N /m^2	
Siever	Sv	1 Sv = 1 J /kg	1 rem = 0.01 Sv
Steradiant	Sr	1 Sr = 1 m^2 m^2	
Watt	W	1 W = 1 J/s	

Conversion of British and American Units to Metric Units

Units of length	m	in	ft	yd	Stat..mile	n. mile
1 m	1	39.3701	3.28084	1.09361	0.00062	0.00054
1 inch	0.0254	1	0.08333	0.02778	-	-
1 foot	0.3048	12	1	0.3333	0.000189	-
1 yard	0.9144	36	3	1	0.0005688	-
1 statute mile	1609.3	63360	5280	1760	1	0.868976
1 n mile	1852	72960	6076.12	2025.12	1.15078	1

Yard, British unit 1 a = 1 are =100 m^2 1 US gallon = 0.0037854 m^2

1 statue mile= 1 Land mile 1 ha =1 hectare = 1000 m^2 1 UK gallon = 0.0045461 m^3

1 n mile = 1 nautical mile; 1 acre = 4,046.86 m^2 1 Barrel petroleum = 0.158971 m^3

1 fathom = 6 ft = 1.8288m 1 US fluid ounce= 1.85 m^3

1 UK fluid ounce= 1.174 mJ

Units of Mass

Units of Mass Weight	Kg	T	Oz	Lb	Sh cwt	Cwt	Sh tn	Ton
1 kg	1	0.001	35.274	2.20462	–	–	–	–
1 t	1000	1	35274	2204.62	22.0462	19.685	1.10231	0.98421
1 oz	0.028535	–	1	0.0625	–	–	–	–
1 lb	0.45359	–	16	1	0.01	0.0089	0.0005	–
1 sh cwt	45.3592	–	–	100	1	0.8929	0.05	0.04464
1 cwt	50.8023	–	–	112	1.12	1	0.056	0.05
1 sh tn	907.185	–	–	2000	20	17.857	1	0.9929
1 ton	1016.05	1.01605	–	2240	22.4	20	1.12	1

T = metric ton = tonne
1 oz = 1 ounce avoirdupois
1 lb = 1 pound avoirdupois

1 sh cwt = 1 short hundred weight, (US unit) = ton (British Unit)
1 cwt = 1 hundred weight, (British unit)
1 short ton (US unit)

Units of Force	N	Dyne	P	Kp	Lbf
1 N	1	105	101.9716	0.10119716	0.224809
1 dyne	10^{-3}	1	1.01976×10^{-3}	1.019710×10^{-6}	2.2489×10^{-3}
1 P	9.80665×10^{-3}	980.665	1	0.001	2.20462×10^{-3}
1 kp	9.80665	9.865×10^{3}	1000	1	2.20462
1 lbf	4.44822	4.44822×10^{5}	453.592	0.453592	1

Dyne (unit of force in the centimeter-gram-second system)
1 kp = 1 kilopound   1 p = 1 pound   1 lbf = 1 pound – force =1 lb x 9.81 m/s^2

Units of Pressure	Pisa	Bar	At	arm	Torr	Psig
1 P 1	1	10^3	$1,01976 \times 10^{-5}$	$0,98692 \times 10^{-2}$	$0,750062 \times 10^{-3}$	$145,038 \times 10^{-8}$
A bar	10^5	1	1,019716	0,986923	750,062	14,5038
1 at	$0,980665 \times 10^5$	0,980665	1	0,967841	735,559	14,2233
1 atm	101,325	1,01325	1,033327	14	760	14,69596
1 Torr	133,3224	$1,333224 \times 10^3$	$1,315789 \times 10^3$	$1,315789 \times 10^3$	1	$19,3368 \times 10^{-3}$
1 psi	$6,89476 \times 10^{-5}$	$68,947 \times 10^{-3}$	703070×10^{-3}	$68,0460 \times 10^{-3}$	51,7128	1

1 bar= 108 dyne/cm^2
1 at = 1 kp/cm^2 = 10 mWC (technical atmosphere)
1 atm = 760 torr(standard atmosphere)
1 psi =1 lbf/in2 (pound-force per sq. inch)

Psia = Pounds per square inch absolute
psig = pounds per square inch, gauge
psid = pounds per square inch differential

Units of Energy	J	kWh	PSh	kpm	Kcal	Btu	SKE
1J =1WS	1	$2,7.78\times10^{-7}$	$3,77\times10^{-7}$	1.1019716	2.388×10^{-4}	9.478×10^{-4}	34.12×10^{-4}
1 kWh	3.6×10^6	1	1,35962	3.671×10^5	859.845	3,412.14	12.28×10^{-2}
1 PSh	2.64×10^6	0,735499	1	2.7×10^5	632.41	2,509.62	90.36×10^3
1 kpm	9,800665	$2,724\times10^{-6}$	$3,70.10^{-6}$	1	2.342×10^{-3}	9.295×10^{-3}	33.47×10^{-8}
1 kcal	4,186.8	$l,163\times10^{-3}$	1.581×10^{-6}	426.935	1	3.96832	14.29×10^{-5}
1 Btu	1,055.06	$2,931\times10^{-4}$	3.985×10^6	107.586	0.251996	1	35.99×10^{-6}
1 SkE	$29,307\times10^{-6}$	8.141	11,067	2.988×10^6	7000	27.78×10^6	1

1 kWe = 1 kilowatt-hour   1 kpm = 1 kilopound-meter

2 BTU = 1 British Thermal Unit   1 PSh = PS-hour

1 kcal = 1 kilocalorie   1 SKE = 1 CE (coal equivalent)

Units of Power	KW	PS	hp	Kpm/s	Kcal/s	Btu/s	Ft-lbf/s
1 KW	1	1.35962	1.34102	101.9716	0.238846	0.94781	737.562
1 PS	0.735499	1	0.986320	75	0.1757	0.69712	542.476
1 hp	0.745700	1:01387	1	76.042	0.17811	0.70679	550
1 kpm/s	9.80×10^{-3}	0.013333	0.0131509	1	2.342×10^{-3}	9.295×10^{-3}	7.23301
1 kcal/s	4.1868	5.692	5.614	426.939	1	3.96832	3.088.05
1 Btu/s	1.05505	1.4345	1.4149	107.586	0.251993	1	778.17
1 ft- lbf/s	1.356×10^{-3}	1.843×10^{-3}	1.818×10^{-3}	0.138255	3.238×10^{-4}	3.285×10^{-3}	1

1 KW   = 1 kilowatt = 10 10 erg/s = 1 kj/s   1 kpm/s = 1 kilopoundmeter per sec.

1 Btu/s = 1 British thermal Unit/sec   1 kcal/s = 1 kilokalorie per sec.

1 PS   = 1 metric horsepower

1 ft-lbf/s = 1 foot-pound / force/sec

1 hp   = 1 horsepower

Source: Siemens AG Pocket Diary, 1998.

☐☐☐

Annexure-V

1. MAJOR PETROLEUM EXPORTING COUNTRIES	2. NUCLEAR SUPPLIERS GROUP COUNTRIES
Algeria	Argentina, Australia, Austria
Bahrain, Brunei Darussalam	Belgium, Brazil, Bulgaria
Congo, China	Canada, Croatia, Czech Republic, Cyprus,
Ecuador	Denmark, Estonia,
Gabon	France, Finland
Indonesia, Iran, Iraq	Germany, Greece, Hungary
Kuwait, Libya	Ireland, Italy, Japan, Rep. of Korea
Nigeria, Oman, Qatar	Kazakistan, Malta, Latvia
Saudi Arabia, Syria	Netherlands, New Zealand, Norway
Trinidad and Tobago	Poland, Portugal, Romania
United Arab Emirates	Russian Federation, Slovakia
Venezuela	Slovenia, South Africa, Spain Sweden, Switzerland, Turkey, Ukrain, United Kingdom United States of America

Annexure-V(A)

NUCLEAR POWER GENERATION CAPACITY IN DIFFERENT COUNTRIES

SL. No.	Country	Number of Reactors Operating (2016)	Installed Capacity (MWe)	% Share in Total National Generation Capacity
1.	Argentina	3	1632	5.6
2.	Armenia	1	375	31.4
3.	Belgium	7	5913	51.7
4.	Brazil	2	1884	2.9
5.	Bulgaria	2	1926	35.0
6.	Canada	19	13554	15.6
7.	China	36	31384	3.6
8.	Czech Republic	6	3930	29.4
9.	Finland	4	2764	33.7
10.	France	58	63130	72.3
11.	Germany	8	10799	13.1
12.	Hungary	4	1889	51.3
13.	India	22	6780	3.4
14.	Iran	1	915	2.1
15.	Japan	43	40290	13.0
16.	S.Korea	25	23077	30.3
17.	Netherlands	1	482	3.4
18.	Mexico	2	1552	6.2
19.	Pakistan	4	1005	4.4
20.	Romania	2	1306	17.1
21.	Russia	37	26528	17.1
22.	Slovakia	4	1814	54.1
23.	Slovenia	1	688	35.2
24.	South Africa	2	1860	6.6
25.	Spain	7	7121	21.4
26.	Sweden	10	9740	40.4
27.	Switzerland	5	3333	34.4
28.	Taiwan	6	5052	13.7
29.	Ukraine	15	13107	52.3
30.	United Kingdom	15	8918	20.4
31.	United States	100	100351	19.7
	World Total	**452**	**392553 MWe**	**10.9**

Annexure-VI

LIST OF BOOKS FOR SUPPLEMENTARY READING

S. No.	Title of the Book	Name of the Author and Publisher
1.	Physics Text Books -Part-1 & Part-2	NCERT, New Delhi, prescribed. (Class-x, ix,xii).
2.	Elements of Electrical and Mechanical Engineering	Dr, S.L.Uppal, P.L.Ballaney, B.D.Indu; Khanna Publishers, Delhi-110006;
3.	An Integrated Course in Electrical Engineering	-Ashok Raj, Harish C Rai, Atam P Dhawan; Umesh Publications, ND.
4.	AEG Manual: Generation,Transmission, Electrical Installations.	AEG, Essen, Germany;
5.	Electrical Technology	H. Cotton; McGraw Hill, USA
6.	Modern Power Station Practices (VoL-.A to H; J to M).	Published by British Electricity International, London
7.	Power Plant Engineering	B. Vijaya Ramnath, C. Elanchezhian, I. Saravanakumr; I.K. International Publishing House
8.	Elements of Nuclear Engineering	Murphy, Glenn, McGraw Hill, USA
9.	Nuclear Reactor Materials and Applicaions	Benjamin M, Deptt. of Nuclear Engg., Lowa State University, USA.
10.	A Text Book of Heat	A W Barton, Longmans, London
11.	Transmission and Distribution	H. Cotton; McGraw Hill, USA
12	.Power Plant Engineering-Black & Veatch	Larry Dabal, Kaya Westra, Pt. Boston; Springer Science & Business Media.
13.	Hydro-Electric Power Station Design	Badger Ralph H, Grant Roy, Nichols HW; Nabu Press.
14.	Hydro-Plant Electrical Systems	David M Clemen; HCI Publication
15.	Nuclear Power Plant Engineering	James H Rust (Amazon)
16.	Nuclear Reactor Engineering	Samuel Glasstone, Alexander Sesonske; Springer Science and Business Media
17.	Power Boilers	John Mackay, James T Pillow (Amazon)
18.	Practical Boiler Operations Engineering and Power Plant.	Amiya Ranjan Mallick, (Amazon)
19.	Thermo-Dynamics and Heat Engines	(Amazon)

Index

Printed in the United States
by Baker & Taylor Publisher Services